D0805814

DISCARD

MASS SPECTROMETRY OF PRIORITY POLLUTANTS

MASS SPECTROMETRY OF PRIORITY POLLUTANTS

Brian S. Middleditch
Stephen R. Missler
and
Harry B. Hines

University of Houston
Houston, Texas

PLENUM PRESS • NEW YORK AND LONDON

Library of Congress Cataloging in Publication Data

Middleditch, Brian S
Mass spectrometry of priority pollutants.

Includes indexes.
1. Organic water pollutants–Analysis. 2. Organic water pollutants–Spectra. 3. Mass spectrometry. I. Missler, Stephen R., joint author. II. Hines, Harry B., joint author. III. Title.
TD427.O7M53 628.1'61 80-14953
ISBN 0-306-40505-9

A Division of Plenum Publishing Corporation
227 West 17th Street, New York, N.Y. 10011

Printed in the United States of America

To
Tin Tin
Betty
Lou-Beth

PREFACE

When the list of organic priority pollutants was first published, many mass spectroscopists went scrambling to their reference books. GC–MS was mandated for the analysis of 114 compounds, yet the spectra of many of them, if they had been recorded at all, were scattered throughout the literature. Moreover, it soon became apparent that, even if a sufficient number of instruments could be made available to undertake the task of monitoring 114 substances in the effluents of 21 categories of industry, the personnel could not be trained to perform the analyses and interpret the results.

The solution to this problem has been the development of highly automated mass spectrometers which can be operated by personnel without the traditional research training. This book is for the new breed of mass spectroscopist who is not interested in the esoteric details of mass spectral fragmentation, but who merely wishes to identify specific pollutants in effluents. Our inclusion of comprehensive lists of synonyms and bibliographic data should make the book of even greater value to the reader who is not too familiar with the idiosyncrasies of chemical nomenclature and the scientific literature. The experienced mass spectroscopist should also benefit from having all of the data collected together in one volume.

This is a book to be used, rather than deposited in a library distant from the laboratory: we would hope that it will find a place on top of every mass spectrometer used for the analysis of priority pollutants.

We are indebted to Pamela Milton for typing the manuscript. Ellis Rosenberg and Betty Bruhns of Plenum have performed an admirable job of implementing the publication of the book. Our wives, to whom this volume is dedicated, have provided the additional sacrifice, support, and encouragement needed for the prompt completion of the manuscript.

Houston

Brian S. Middleditch
Stephen R. Missler
Harry B. Hines

CONTENTS

INTRODUCTION

The Federal Water Pollution Control Act Amendments of 1972 (P.L. 92-500) required the U.S. Environmental Protection Agency to develop a comprehensive program to improve the quality of the nation's waterways. Section 307(a) mandated publication of a list of toxic pollutants for which effluent standards were being established. There was substantial delay in implementing this task. It was not until June 7, 1976 that the National Resources Defense Council, Environmental Defense Fund, Businessmen for the Public Interest, National Audubon Society, and Citizens for a Better Environment were able to reach an agreement with the EPA under which timetables and procedures for implementation of certain sections of P.L. 92-500 were delineated. One of the terms of the consent decree related to the study of 65 substances or groups of substances (totaling 129 individual substances) in industrial effluents, and the development of appropriate regulations for their control. These substances (known as "priority pollutants," "consent decree pollutants," or "toxic pollutants") were selected on the basis of their known occurrence in effluents, their presence in drinking water or fish, their known or suspected carcinogenic, mutagenic, or teratogenic properties, their likelihood of human exposure, their persistence in the aquatic food web, their propensity for bioaccumulation, and their toxicity to aquatic organisms and those (including humans) which might feed on such organisms.

Combined gas chromatography–mass spectrometry (GC–MS) was required for the analysis of the 114 organic priority pollutants, according to guidelines for sampling and analysis distributed by the EPA in early 1977.

There are four categories of organic priority pollutants (volatiles, base–neutral extractables, acid extractables, and pesticides), each requiring a different analytical procedure.

The volatile priority pollutants are listed in Table 1. They are analyzed by the purge-and-trap method. The pollutants are purged from solution by helium onto a trap containing Tenax-GC and silica gel. After an appropriate purging time, the trapped compounds are desorbed onto a GC column for analysis by GC–MS. Bromochloromethane, 2-bromo-1-chloropropane, and 1,4-dichlorobutane are employed as internal standards. There is some overlap between the compounds that can be analyzed by the purge-and-trap method and those that can be analyzed by liquid–liquid extraction methods. Some of the compounds listed in Table 1 are not easily determined by the purge-and-trap method; these are noted in the text.

The base–neutral extractable priority pollutants are listed in Table 2. The sample is adjusted to pH 11 using sodium hydroxide, and extraction with

Table 1. Volatile Priority Pollutants

Acrolein	1,2-Dichloroethane
Acrylonitrile	1,1-Dichloroethylene
Benzene	*trans*-1,2-Dichloroethylene
Bis(chloromethyl) ether	1,2-Dichloropropane
Bromodichloromethane	*cis*-1,3-Dichloropropene
Bromoform	*trans*-1,3-Dichloropropene
Bromomethane	Ethylbenzene
Carbon tetrachloride	Methylene chloride
Chlorobenzene	1,1,2,2-Tetrachloroethane
Chloroethane	1,1,2,2-Tetrachloroethene
2-Chloroethyl vinyl ether	Toluene
Chloroform	1,1,1-Trichloroethane
Chloromethane	1,1,2-Trichloroethane
Dibromochloromethane	Trichloroethylene
Dichlorodifluoromethane	Trichlorofluoromethane
1,1-Dichloroethane	Vinyl chloride

Table 2. Base-Neutral Extractable Priority Pollutants

Acenaphthene	Diethyl phthalate
Acenaphthylene	Dimethyl phthalate
Anthracene	2,4-Dinitrotoluene
Benzidine	2,6-Dinitrotoluene
Benzo[*a*]anthracene	Di-n-octyl phthalate
Benzo[*b*]fluoranthene	1,2-Diphenylhydrazine
Benzo[*k*]fluoranthene	Fluoranthene
Benzo[*ghi*]perylene	Fluorene
Benzo[*a*]pyrene	Hexachlorobenzene
Bis(2-chloroethoxy)methane	Hexachlorobutadiene
Bis(2-chloroethyl) ether	Hexachlorocyclopentadiene
Bis(2-chloroisopropyl) ether	Hexachloroethane
Bis(2-ethylhexyl) phthalate	Indeno[1,2,3-*cd*]pyrene
4-Bromophenyl phenyl ether	Isophorone
Butyl benzyl phthalate	Naphthalene
2-Chloronaphthalene	Nitrobenzene
4-Chlorophenyl phenyl ether	*N*-Nitrosodimethylamine
Chrysene	*N*-Nitrosodiphenylamine
Dibenzo[*a*,*h*]anthracene	*N*-Nitrosodi-n-propylamine
Di-n-butyl phthalate	Phenanthrene
1,2-Dichlorobenzene	Pyrene
1,3-Dichlorobenzene	2,3,7,8-Tetrachlorodibenzo-*p*-dioxin
1,4-Dichlorobenzene	1,2,4-Trichlorobenzene
3,3′-Dichlorobenzidine	

Table 3. Acid Extractable Priority Pollutants[a]

p-Chloro-*m*-cresol	2-Nitrophenol
2-Chlorophenol	4-Nitrophenol
2,4-Dichlorophenol	Pentachlorophenol
2,4-Dimethylphenol	Phenol
4,6-Dinitro-*o*-cresol	2,4,6-Trichlorophenol
2,4-Dinitrophenol	

[a]"Total phenols" are also to be measured.

methylene chloride affords these compounds. They are analyzed by GC–MS, using anthracene-d_{10} as an internal standard.

The acid extractable priority pollutants are listed in Table 3. After extracting the base-neutrals, the remaining water sample is adjusted to pH 2 using hydrochloric acid. Extraction with methylene chloride affords the phenols. They are analyzed by GC–MS, also using anthracene-d_{10} as an internal standard. Some of the phenols do not possess good gas chromatographic properties.

The pesticide priority pollutants are listed in Table 4. The PCBs are included in this category even though they are not pesticides. These compounds are analyzed by gas chromatography with electron capture detection. If pesticides are detected, their identities are confirmed by GC–MS.

Table 4. Pesticide Priority Pollutants

Aldrin	Dieldrin	PCB-1016[a]
α-BHC	α-Endosulfan	PCB-1221[a]
β-BHC	β-Endosulfan	PCB-1232[a]
γ-BHC	Endosulfan sulfate	PCB-1242[a]
δ-BHC	Endrin	PCB-1248[a]
Chlordane[a]	Endrin aldehyde	PCB-1254[a]
4,4′-DDD	Heptachlor	PCB-1260[a]
4,4′-DDE	Heptachlor epoxide	Toxaphene[a]
4,4′-DDT		

[a]These substances are mixtures.

Table 5. Inorganic Priority Pollutants[a]

Antimony	Chromium	Nickel
Arsenic	Copper	Selenium
Asbestos	Cyanide	Silver
Beryllium	Lead	Thallium
Cadmium	Mercury	Zinc

[a]This category includes salts of each of these substances (except asbestos). Ammonia has also been proposed as a member of this category.

The inorganic priority pollutants are listed in Table 5. Mass spectrometry is not used in their analysis.

Complete details of sampling and analysis procedures are obtainable from the EPA Environmental Monitoring and Support Laboratory, Cincinnati, Ohio 45268, so they are not repeated here.

Data Compilations

Our data compilations contain mass spectra (line diagrams and tabulated data), lists of synonyms, brief notes on analyses and uses, selected bibliographies, and additional information useful in searching the literature for further data on each compound.

The information given is listed in the order normally used by the EPA. The EPA nomenclature has been adhered to even though, in some instances, it deviates from the IUPAC or CAS conventions. For each substance or group of substances the following data are given.

Class of Pollutant. This is the description of the pollutant (or group of pollutants) as stated in the consent decree. An indication is given of the use (if any) of each individual substance. Some compounds, particularly the halogenated compounds, are formed during chlorination of drinking water, although this is not always mentioned. Salient details of analytical procedures or mass spectra are given where warranted.

Individual Pollutant. The EPA name, molecular formula, and molecular weight (calculated using integral atomic weights of most abundant isomers, as is conventional in mass spectrometry) are given.

The category of pollutant follows.

The name assigned for the *Eighth Collective Index* of Chemical Abstracts (CAS name) and other synonyms are followed by the CAS registry number. The registry number is a unique identifier which is invaluable in searching the literature for further information about each compound.

References are also given to the Registry of Toxic Effects of Chemical Substances (which provides toxicity data) and the ninth edition of the Merck Index (which provides chemical and pharmaceutical data).

Spectral Data. The eight most abundant ions in each spectrum are given, as are the ions considered by the EPA to be characteristic of each compound or useful for quantitation. Mass spectral data are given, in tabulated form and as line diagrams, for ions greater in relative abundance than 2% of the base peak. The minimum m/e value for volatiles is m/e 20, while for nonvolatiles it is m/e 40. Ions of nonintegral mass are not tabulated.

Reference compounds were obtained from Aldrich Chemical Co., Analabs, Inc., Chem-Service, Nanogens Co., RFR Corp., and Supelco, Inc. All of the data were acquired by GC-MS using a Hewlett-Packard 5992A hyperbolic quadrupole mass spectrometer in the authors' laboratory. All quadrupole mass spectrometers will, if not used carefully, produce spectra with varying relative abundances of ions throughout the mass range. This can be monitored using decafluorotriphenylphosphine as a reference compound and the tuning of the instrument adjusted accordingly. If these precautions are taken, spectra obtained by others should be similar to those reported here.

Another source of variation between spectra acquired using different instruments is the ease of formation of multiply charged ions. These ions (or the corresponding isotopic species) are easily recognized if they appear at nonintegral m/e values. In such instances, particular attention should be given to the relative abundances of nearby doubly charged ions of integral mass, since they may not be reproducible.

Selected Bibliography. Up to five citations of the relevant literature are given for each substance. Both Chemical Abstracts and Gas Chromatography-Mass Spectrometry Abstracts were searched.

Comprehensive indexes follow the data compilations.

Selected Bibliography

Bellar, T. A., and Lichtenberg, J. J., Determining volatile organics at microgram-per-liter levels by gas chromatography, *J.A.W.W.A.* **66** (12), 739–744 (1974).

Harris, L. E., Budde, W. L., and Eichelberger, J. W., Direct analysis of water samples for organic pollutants with gas chromatography-mass spectrometry, *Anal. Chem.* **46** (13), 1912–1917 (1974).

Determination of organochlorine pesticides in industrial effluents, *Federal Register* **38** (125), part II, appendix II, 17319 (June 29, 1975).

Eichelberger, J. W., Harris, L. E., and Budde, W. L., Reference compound to calibrate ion abundance measurements in gas chromatography-mass spectrometry systems, *Anal. Chem.* **47**, 995–1000 (1975).

Heller, S. R., McGuire, J. M., and Budde, W. L., Trace organics by GC/MS, *Environ. Sci. Technol.* 9 (3), 210–213 (1975).

Quality Criteria for Water, U.S. E.P.A., Washington, D.C. (1976).

Keith, L. H. (ed.), *Identification and Analysis of Organic Pollutants in Water*, Ann Arbor Science Publishers, Ann Arbor, Michigan (1976).

Sampling and Analysis Procedures for Screening of Industrial Effluents for Priority Pollutants, U.S. E.P.A., Environmental Monitoring and Support Laboratory, Cincinnati, Ohio (1977) (revised).

Mackenthum, K. M., Setting standards for wastewater effluents–present status and future trends, pp. 203–213, in Borchardt, J. A., Cleland, J. K., Redman, W. J., and Olivier, G. (eds.), *Viruses and Trace Contaminants in Water and Wastewater*, Ann Arbor Science Publishers, Ann Arbor, Michigan (1977).

ACENAPHTHENE

This compound is, strictly, not a polynuclear aromatic hydrocarbon since it is not fully unsaturated, but the comments on these compounds (*q.v.*) are pertinent.

Acenaphthene (derived from coal tar) was once used as an insecticide and fungicide. It has also been used in the manufacture of dyes and plastics.

ACENAPHTHENE **$C_{12}H_{10}$ (154)**

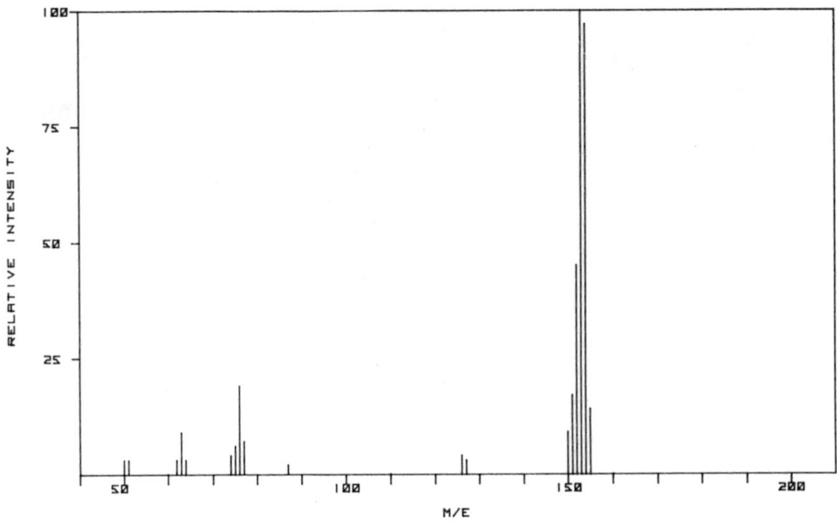

Spectral Data

Mass	Abundance	Mass	Abundance	Mass	Abundance
50	2.6	75	5.9	150	8.5
51	3.4	76	19.0	151	16.7
62	2.9	77	7.3	152	44.6
63	8.9	87	2.3	153	100.0
64	2.5	126	4.3	154	96.5
74	3.9	127	3.4	155	14.0

ACENAPHTHENE–*continued*

Base–neutral extractable
CAS Name: Acenaphthene
Synonyms
1,2-Dihydroacenaphthylene
peri-Ethylenenaphthalene
1,8-Ethylenenaphthalene
Naphthyleneethylene

CAS Registry No.: 83-32-9
Merck Index Ref.: 19

Major Ions: 153, 154, 152, 76, 151, 155, 63, 150
EPA Ions: 154, 153, 152

Selected Bibliography

Data processing of low-energy ionization mass spectra and type determination of aromatic hydrocarbons, Nishishita, T., Yoshihara, M., Oshima, S., *Maruzen Sekiyu Giho*, (16), 99–104 (1971).

Electron impact studies of some cyclic hydrocarbons, Frey, W. F., Compton, R. N., Naff, W. T., Schweinler, H. C., *Int. J. Mass Spectrom. Ion Phys.*, **12**(1), 19–32 (1973).

Mixed charge exchange–chemical ionization mass spectrometry of polycyclic aromatic hydrocarbons, Lee, M. L., Hites, R. A., *J. Am. Chem. Soc.*, **99**(6), 2008–2009 (1977).

Preparation and low-voltage mass spectrometry sensitivities of methylated polynuclear aromatic hydrocarbons, Schiller, J. E., *Anal. Chem.*, **49**(8), 1260–1262 (1977).

ACROLEIN

Although listed as a "volatile," acrolein is insufficiently volatile to be purged from aqueous solution at room temperatures. Recovery is enhanced at elevated temperatures. Direct aqueous injection is suggested as an alternative procedure. Phenylhydrazone derivatives are employed for the colorimetric assay of acrolein, and are also amenable to mass spectrometry.

Acrolein is used for aquatic weed control and also for inhibiting the growth of sulfur-bacteria in oil field brines. Among its other uses are in the production of plastics, perfumes, and colloidal forms of metals. It has been used in chemical warfare agents and as a warning agent in methyl chloride refrigerants.

ACROLEIN **C_3H_4O (56)**

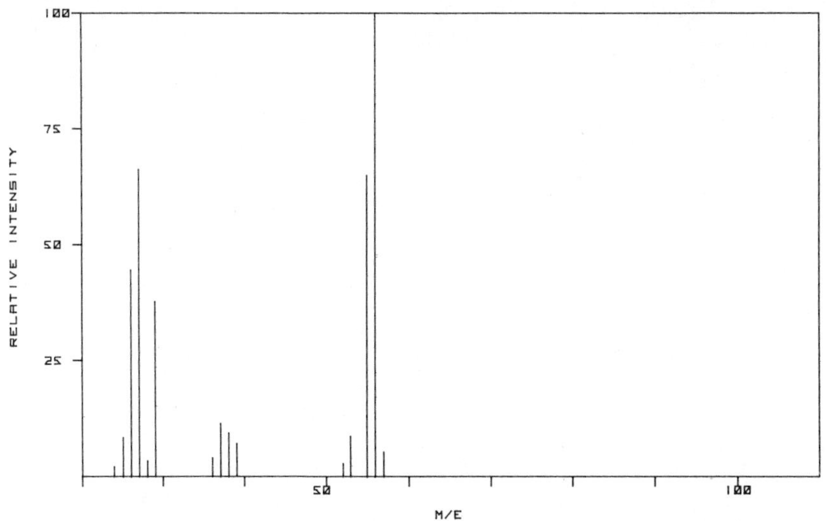

Spectral Data

Mass	Abundance	Mass	Abundance	Mass	Abundance
24	2.1	29	37.7	52	2.6
25	8.3	36	4.0	53	8.6
26	44.5	37	11.3	55	64.9
27	66.2	38	9.3	56	100.0
28	3.3	39	7.1	57	5.3

ACROLEIN–*continued*

Volatile
CAS Name: Acrolein
Synonyms

Acquinite	Aqualin	Propenal
Acraldehyde	Aqualine	2-Propenal
Acrylaldehyde	Biocide	2-Propen-1-one
Acrylic aldehyde	Ethylene aldehyde	Slimicide
Allyl aldehyde	Magnacide H	

CAS Registry No.: 107-02-8 (formerly 25314-61-8)
ROTECS Ref.: AS10500
Merck Index Ref.: 123

Major Ions: 56, 27, 55, 26, 29, 37, 38, 53
EPA Ions: 26, 27, 55, 56

Selected Bibliography

Thin-layer chromatographic separation and determination of the 2,4-dinitrophenylhydrazones of several carbonyl compounds at the nanogram level, Kolbe, M, Seifert, B., *Fresenius' Z. Anal. Chem.*, **281**(5), 365–369 (1976).

ACRYLONITRILE

As with acrolein, purging at room temperature is impractical, but direct aqueous injection may be used.

Acrylonitrile is used as a fumigant and also in the manufacture of acrylic plastics.

ACRYLONITRILE **C_3H_3N (53)**

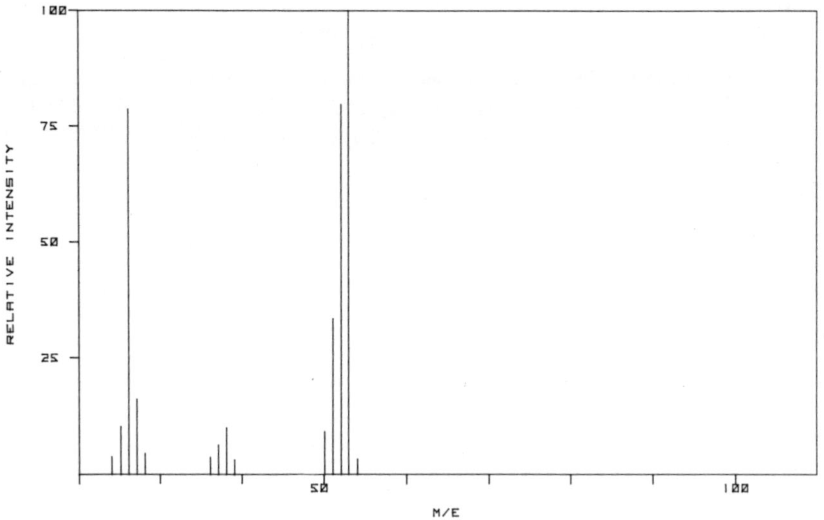

Spectral Data

Mass	Abundance	Mass	Abundance	Mass	Abundance
24	3.7	36	3.6	51	33.5
25	10.2	37	6.2	52	79.6
26	78.5	38	9.8	53	100.0
27	16.0	39	3.0	54	3.2
28	4.4	50	9.1		

Volatile
CAS Name: Acrylonitrile

ACRYLONITRILE–*continued*

Synonyms

Acritet	Fumigrain	2-Propenenitrile
Acrylon	Miller's Fumigrain	TL 314
Carbacryl	NCI-C50215	VCN
Cyanoethylene	Propenenitril	Ventox
Ent 54	Propenenitrile	Vinyl cyanide

CAS Registry No.: 107-13-1 (formerly 29754-21-0, 43690-95-5, 43690-96-6)
ROTECS Ref.: AT52500
Merck Index Ref.: 127

Major Ions: 53, 52, 26, 51, 27, 25, 38, 50
EPA Ions: 26, 51, 52, 53

Selected Bibliography

Measurement of negative ions formed by electron impact. IX. Negative ion mass spectra and ionization efficiency curves of negative ions of *m/e* 25, 26, 27, 38, 39, 40, and 50 from acrylonitrile, Tsuda, S., Yokohata, A., Umaba, T., *Bull. Chem. Soc. Jap.*, **46**(8), 2273–2277 (1973).

BENZENE

As for the polynuclear aromatic hydrocarbons (*q.v.*), benzene affords an abundant molecular ion, few fragment ions, and multiply-charged molecular ions.

Benzene is a widely used solvent and intermediate in the manufacture of many chemicals. It has been used to destroy screwworm larvae.

BENZENE C_6H_6 (78)

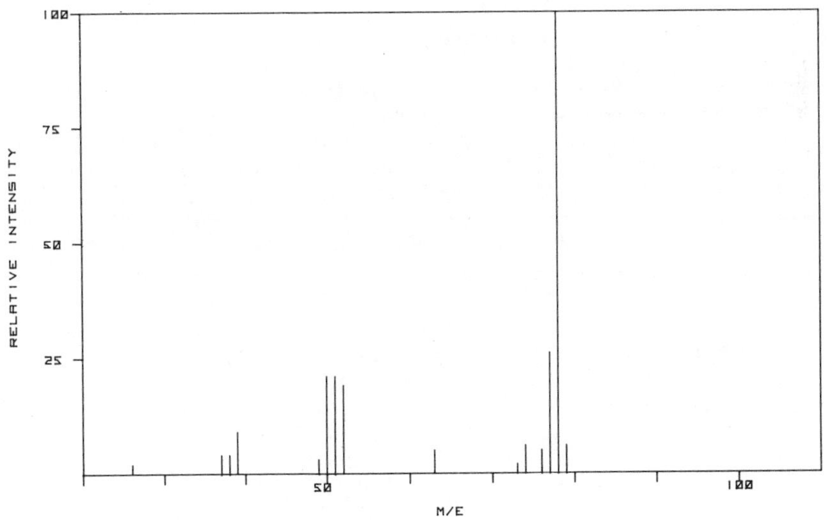

Spectral Data

Mass	Abundance	Mass	Abundance	Mass	Abundance
26	2.1	50	20.8	74	5.8
37	3.5	51	20.6	76	5.4
38	4.2	52	18.8	77	25.9
39	9.3	63	4.5	78	100.0
49	3.3	73	2.1	79	6.4

Volatile
CAS Name: Benzene

BENZENE–*continued*

Synonyms

(6)Annulene	Carbon oil	Nitration benzene
Benzol	Coal naphtha	Phenyl hydride
Benzole	Cyclohexatriene	Pyrobenzol
Benzolene	Mineral naphtha	Pyrobenzole
Bicarburet of hydrogen	Motor benzol	

CAS Registry No.: 71-43-2 (formerly 54682-86-9)
ROTECS Ref.: CY14000
Merck Index Ref.: 1069

Major Ions: 78, 77, 50, 51, 52, 39, 79, 74
EPA Ion: 78

Selected Bibliography

Multiply charged ions in the mass spectra of aromatics, Engel, R., Halpern, D., Funk, B. A., *Org. Mass Spectrom.*, **7**(2), 177–183 (1973).

The collection and analysis of volatile hydrocarbon air pollutants using a timed elution chromatographic technique linked to a computer controlled mass spectrometer, Perry, R., Twibell, J. D., *Biomed. Mass Spectrom.*, **1**(1), 73–77 (1974).

Assessment of the trace organic molecular composition of industrial and municipal wastewater effluents by capillary gas chromatography/real-time high-resolution mass spectrometry: a preliminary report, Burlingame, A. L., *Ecotoxicol. Environ. Saf.*, **1**(1), 111–150 (1977).

A survey of the molecular nature of primary and secondary components of particles in urban air by high-resolution mass spectrometry, Cronn, D. R., Charlson, R. J., Knights, R. L., Crittenden, A. L., Appel, B. R., *Atmos. Environ.*, **11**(10), 929–937 (1977).

A field portable mass spectrometer for monitoring organic vapors, Meier, R. W., *J. Am. Ind. Hyg. Assoc.*, **39**(3), 233–239 (1978).

Computer-controlled gas chromatographic/high accuracy mass spectrometric analysis of organic emissions from stationary sources, Schuetzle, D., Prater, T. J., Harvey, T. M., Hagge, D. E., *Adv. Mass Spectrom.*, **7B**, 1082–1090 (1978).

BENZIDINE

This compound has been recognized as a carcinogen for many years, but has been largely ignored by mass spectroscopists.

Benzidine was formerly used by analytical chemists in the determination of sulfates, hydrogen peroxide, blood, and some metals. It was also used in the manufacture of dyes.

BENZIDINE $C_{12}H_{12}N_2$ (184)

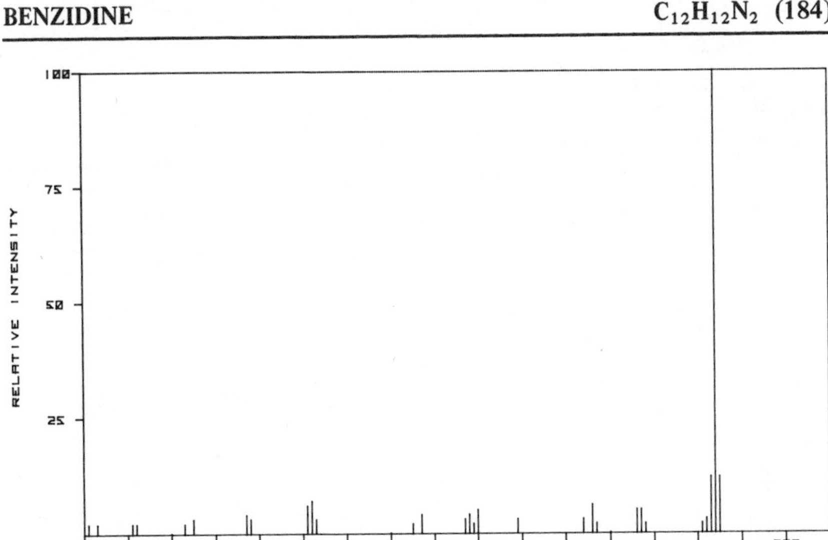

Spectral Data

Mass	Abundance	Mass	Abundance	Mass	Abundance
41	2.3	93	2.7	157	2.2
43	2.0	115	2.3	166	4.9
51	2.0	117	3.8	167	5.1
52	2.4	127	2.9	168	2.1
63	2.2	128	3.9	181	2.1
65	3.0	129	2.1	182	3.1
77	3.6	130	4.6	183	11.5
78	3.1	139	2.5	184	100.0
91	5.9	154	2.6	185	12.3
92	6.8	156	6.4		

BENZIDINE–*continued*

Base–neutral extractable
CAS Name: (1,1′-Biphenyl)-4,4′-diamine
Synonyms

- 4,4′-Biphenyldiamine
- C.I. azoic diazo component 112
- 4,4′-Diaminobiphenyl
- 4,4′-Diaminodiphenyl
- *p*-Diaminodiphenyl
- 4,4′-Diphenylenediamine
- Fast corinth base B

CAS Registry: 92-87-5 (formerly 56481-94-8)
ROTECS Ref.: DC96250
Merck Index Ref.: 1083

Major Ions: 184, 185, 183, 92, 156, 91, 167, 166
EPA Ions: 184, 92, 185

CARBON TETRACHLORIDE

Carbon tetrachloride is registered for use as a grain fumigant. This compound was formerly employed in dry cleaning and in fire extinguishers (for use on electrical fires), but these uses were discontinued when it was found that phosgene was generated by oxidation. Phosgene has recently also been identified as a metabolite of carbon tetrachloride.

Carbon tetrachloride has also been used as an anthelmintic agent (to destroy parasitic worms).

CARBON TETRACHLORIDE **CCl_4 (152)**

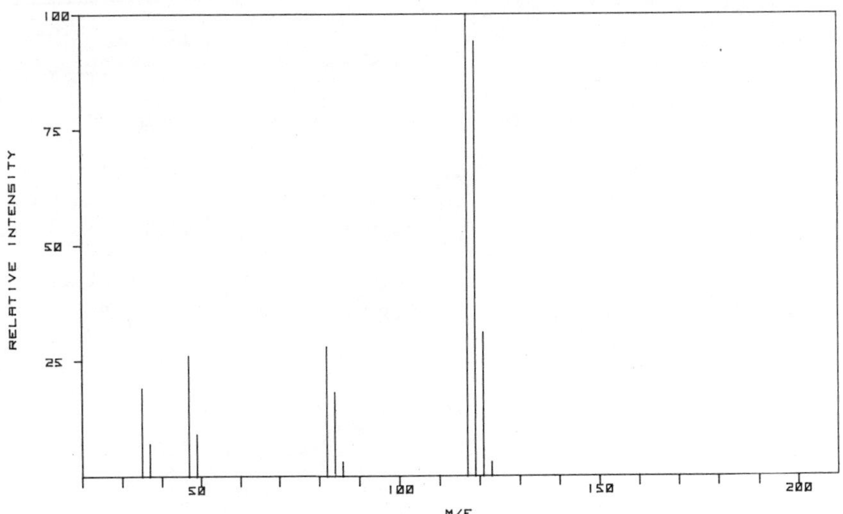

Spectral Data

Mass	Abundance	Mass	Abundance	Mass	Abundance
35	18.7	82	28.3	119	93.7
37	6.6	84	18.3	121	30.7
47	25.6	86	2.9	123	3.2
49	8.9	117	100.0		

CARBON TETRACHLORIDE–*continued*

Volatile
CAS Name: Methane, tetrachloro-
Synonyms

Benzinoform	Flukoids	Necatorina
Carbona	Freon 10	Necatorine
Carbon chloride	Halon 104	Perchloromethane
Ent 4,705	Methane tetrachloride	Tetrachloromethane

CAS Registry No.: 56-23-5
ROTECS Ref.: FG49000
Merck Index Ref.: 1821

Major Ions: 117, 119, 121, 82, 47, 35, 84, 49
EPA Ions: 117, 119, 121

Selected Bibliography

Positive-ion mass spectra of some perhalogenated compounds, Contineanu, M. A., Grubel, K., *An. Univ. Bucuresti, Chim.*, **20**(2), 175–181 (1971).

Azimuthal partial dipole moment analysis of the mass spectra of methane and its chlorine derivatives, Tsai, Tsu-Min, Tchen, Saw-Chow, Chen, Cheng, *Hua Hsueh*, (4), 111–116 (1973).

Direct aqueous injection gas chromatography–mass spectrometry for analysis of organohalides in water at concentrations below the parts per billion level, Fujii, T. *J. Chromatogr.*, **139**(2), 297–302 (1977).

Head space mass spectrometric analysis for volatiles in biological specimens, Urich, R. W., Bowerman, D. L., Wittenberg, P. H., Pierce, A. F., Schisler, D. K., Levisky, J. A., Pflug, J. L., *J. Anal. Toxicol.*, **1**(5), 195–199 (1977).

Chlorine-35/chlorine-37 isotope effects in the fragmentations of metastable perchlorinated alkane and silane ions, Potzinger, P., Basu, S., *Ber. Bunsenges. Phys. Chem.*, **82**(4), 415–418 (1978).

CHLORINATED BENZENES
(other than dichlorobenzenes)

Chlorobenzene, hexachlorobenzene, and 1,2,4-trichlorobenzene are included in this category. The 1,2-, 1,3-, and 1,4-dichlorobenzenes are listed under the heading "dichlorobenzenes" (*q.v.*).

Chlorobenzene is used in the manufacture of DDT, aniline, and phenol. It is also a paint solvent and a heat transfer agent.

Hexachlorobenzene is a fungicide used for treatment of wheat seed against bunt (a smut of *Tilletia foetida* or *T. caries*). Accidental use of treated seed in the production of flour in Turkey led to widespread poisoning (cutaneous porphyria).

A technical preparation of 1,3,5-trichlorobenzene (TCB, TCBA, Polystream) used for termite control contains other isomers and 1,2,3,4-tetrachlorobenzene.

CHLOROBENZENE C_6H_5Cl (112)

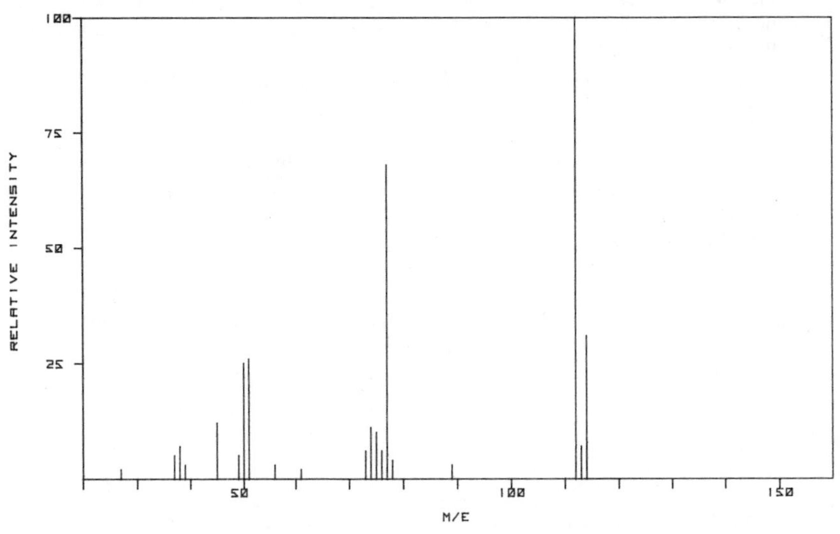

CHLOROBENZENE–*continued*

Spectral Data

Mass	Abundance	Mass	Abundance	Mass	Abundance
27	2.1	51	25.8	77	67.6
37	5.2	56	3.1	78	4.4
38	7.3	61	2.2	89	3.2
39	2.6	73	6.1	112	100.0
45	11.9	74	11.4	113	7.0
49	4.6	75	9.7	114	31.4
50	25.0	76	6.2		

Volatile
CAS Name: Benzene, chloro-
Synonyms

Benzene chloride	Chlorobenzol	Monochlorobenzene
Chlorbenzene	MCB	NCI-C54886
Chlorbenzol	Monochlorbenzene	Phenyl chloride

CAS Registry No.: 108-90-7 (formerly 50717-45-8)
ROTECS Ref.: CZ01750
Merck Ref.: 2095

Major Ions: 112, 77, 114, 51, 50, 45, 74, 75
EPA Ions: 112, 114

Selected Bibliography

Fragmentation diagrams for the elucidation of decomposition reactions of organic compounds. II. Polymethyl- and chlorobenzenes, Schoenfeld, W., *Org. Mass Spectrom.*, **10**(6), 401–426 (1975).

Structural and energetics effects in the chemical ionization of halogen-substituted benzenes and toluenes, Leung, Hei-Wun, Harrison, A. G., *Can. J. Chem.*, **54**(21), 3439–3452 (1976).

Ionization and fragmentation in mass spectrometry. VI. Activation energy of the principal fragmentation of monohalobenzenes, Bouchoux, G., *Org. Mass Spectrom.*, **12**(11), 681–684 (1977).

A computer algorithm for qualitative identification of mass spectral data acquired in trace level analysis of environmental samples, Davidson, W. C., Smith, M. J., Schaefer, D. J., *Anal. Lett.*, **10**(4), 309–331 (1977).

Mass spectrometric investigations of doubly charged ions by field ionization, Migahed, M. D., Helal, A. I., *Egypt. J. Phys.*, 7(1), 41–46 (1977).

HEXACHLOROBENZENE C_6Cl_6 (282)

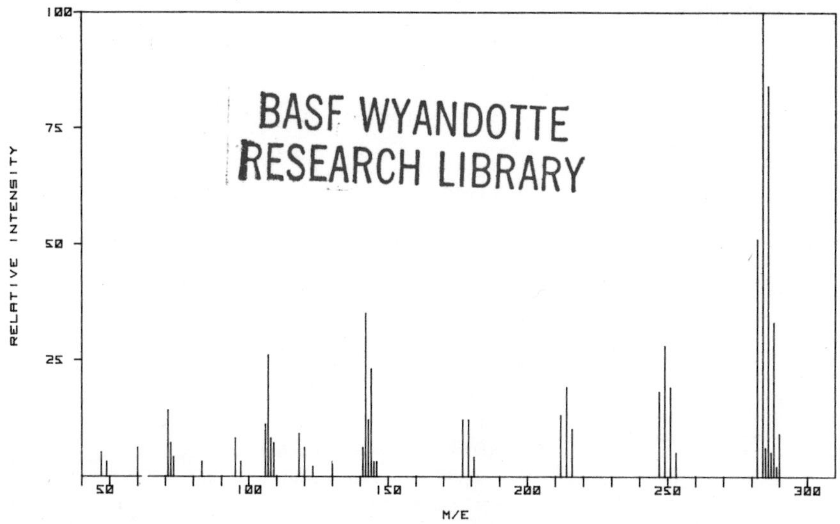

Spectral Data

Mass	Abundance	Mass	Abundance	Mass	Abundance
47	4.7	120	5.5	216	9.9
49	3.0	123	2.2	247	17.6
60	5.8	130	3.0	249	28.1
71	13.5	141	5.8	251	18.5
72	6.6	142	35.3	253	5.2
73	4.4	143	12.1	282	50.7
83	3.0	144	22.6	284	100.0
95	7.7	145	2.5	285	6.1
97	3.0	146	3.0	286	83.7
106	10.7	177	12.1	287	5.2
107	25.6	179	11.8	288	32.8
108	8.0	181	3.9	289	2.2
109	7.4	212	13.2	290	8.8
118	8.5	214	19.3		

Base-neutral extractable
CAS Name: Benzene, hexachloro-
Synonyms

Amatin	Hexa C. B.	Pentachlorophenyl chloride
Anticarie	Julin's carbon chloride	Perchlorobenzene
Bunt-cure	No Bunt	Phenyl perchloryl
Bunt-no-more	No Bunt 40	Sanocide
Co-op Hexa	No Bunt 80	Smut-go
HCB	No Bunt Liquid	Snieciotox

HEXACHLOROBENZENE–*continued*

CAS Registry No.: 118-74-1
ROTECS Ref.: DA29750
Merck Index Ref: 4544

Major Ions: 284, 286, 282, 142, 288, 249, 107, 144
EPA Ions: 284, 142, 249

Selected Bibliography

Mass spectra of some perchloroaromatic compounds, Rivera, J., Castaner, J., Ballester, M., *An. Quim.*, **70**(12), 1194–1198 (1974).

Mass-spectrometric identification and quantitation of chlorinated hydrocarbons in fish, Schaefer, R. G., *Chem.-Ztg.*, **98**(5), 241–247 (1974).

Assessment of the trace organic molecular composition of industrial and municipal wastewater effluents by capillary gas chromatography/real-time high-resolution mass spectrometry: a preliminary report, Burlingame, A. L., *Ecotoxicol. Environ. Saf.*, **1**(1), 111–150 (1977).

Application of coupled gas chromatography–mass spectrometry in methods for the study and determination of pesticide residues and micropollutants in environmental and food materials, Mestres, R., Chevallier, C., Espinoza, C., Cornet, R., *Ann. Falsif. Exper. Chim.*, **70**(751), 177–188 (1977).

1,2,4-TRICHLOROBENZENE $C_6H_3Cl_3$ (180)

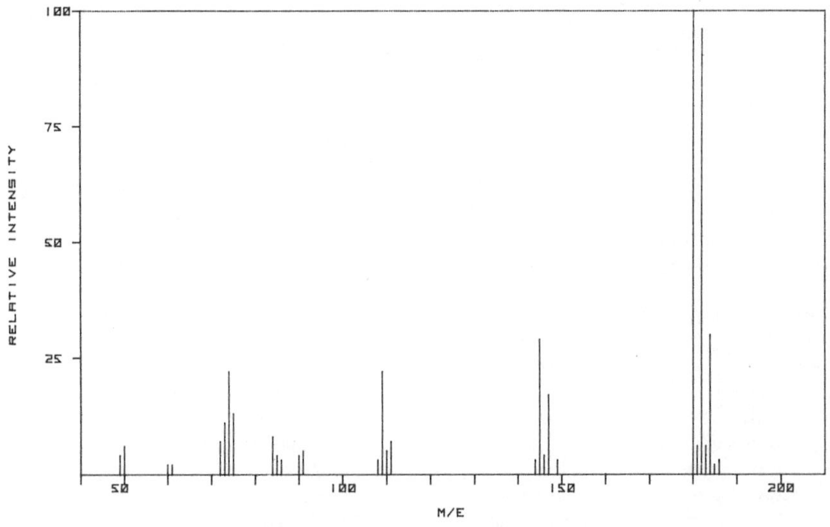

Spectral Data

Mass	Abundance	Mass	Abundance	Mass	Abundance
49	3.8	86	2.9	147	17.1
50	6.2	90	4.2	149	2.8
60	2.3	91	4.7	180	100.0
61	2.3	108	3.4	181	6.3
72	6.5	109	22.1	182	96.3
73	11.4	110	5.2	183	6.1
74	21.6	111	7.4	184	30.2
75	12.8	144	3.0	185	2.1
84	8.4	145	29.0	186	3.0
85	3.8	146	3.6		

Base–neutral extractable
CAS Name: Benzene, 1,2,4-trichloro-
Synonyms

unsym-Trichlorobenzene
1,2,5-Trichlorobenzene
1,2,4-Trichlorobenzol

CAS Registry No.: 120-82-1
ROTECS Ref.: DC21000
Merck Index Ref.: 9310

Major Ions: 180, 182, 184, 145, 109, 74, 147, 75
EPA Ions: 180, 182, 145

CHLORINATED ETHANES

Chloroethane, 1,1-dichloroethane, 1,2-dichloroethane, hexachloroethane, 1,1,-2,2-tetrachloroethane, 1,1,1-trichloroethane, and 1,1,2-trichloroethane are included in this category.

Chloroethane is used as a refrigerant and in the manufacture of tetraethyl lead. It is also used as a topical anesthetic, a solvent, and an alkylating agent.

1,2-Dichloroethane is a fumigant and solvent.

Hexachloroethane is a solvent and an anthelmintic agent (flukicide). It is also used in rubber vulcanization and in celluloid manufacture.

1,1,2,2-Tetrachloroethane is also a solvent and an anthelmintic agent (hookworms and trematodes). Other uses are in dry cleaning and metal degreasing.

1,1,1-Trichloroethane has been used as a fumigant and in the cleaning of plastic molds and cold type metal.

1,1,2-Trichloroethane is a solvent.

CHLOROETHANE C_2H_5Cl (64)

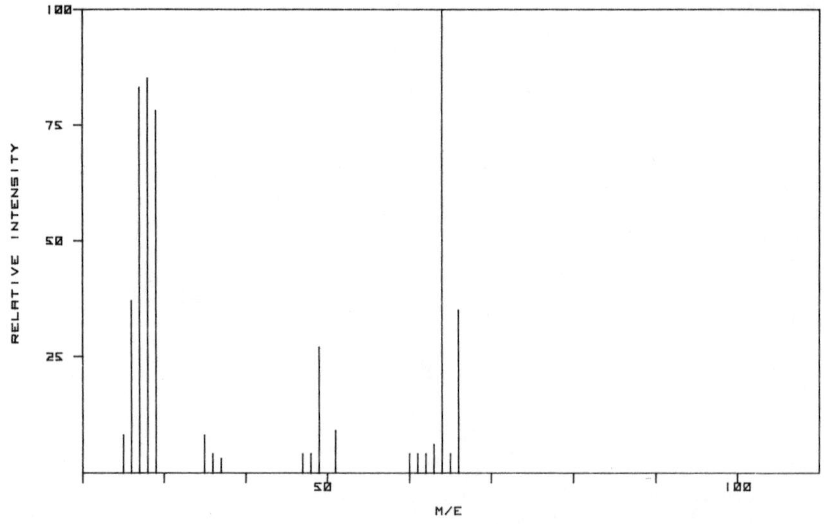

Spectral Data

Mass	Abundance	Mass	Abundance	Mass	Abundance
25	7.6	37	3.1	61	4.0
26	36.9	47	4.3	62	3.7
27	82.7	48	4.1	63	6.3
28	84.9	49	27.0	64	100.0
29	78.1	51	8.8	65	4.2
35	7.6	60	3.8	66	35.0
36	4.4				

Volatile
CAS Name: Ethane, chloro-
Synonyms

Aethylis	Chloryl anesthetic
Aethylis chloridum	Ether muriatic
Anodynon	Ethyl chloride
Chelen	Hydrochloric ether
Chlorethyl	Kelene
Chloridum	Monochloroethane
Chloroaethan	Muriatic ether
Chloryl	Narcotile

CAS Registry No.: 75-00-3
ROTECS Ref.: KH75250
Merck Index Ref.: 3713

Major Ions: 64, 28, 27, 29, 26, 66, 49, 51
EPA Ions: 64, 66

Selected Bibliography

Kinetic energies of fragment ions from some hydrocarbons and organic halides in a modified mass spectrometer, Ossinger, A. I., Weiner, E. R., *J. Chem. Phys.*, **65**(7), 2892–2900 (1976).

A high-pressure mass-spectrometric study of ion–molecule reactions in ethyl fluoride and ethyl chloride, Tan, H. S., Pabst, M. J. K., Franklin, J. L., *Int. J. Mass Spectrom. Ion Phys.*, **21**(3–4), 297–315 (1976).

A computer algorithm for qualitative identification of mass spectral data acquired in trace level analysis of environmental samples, Davidson, W. C., Smith, M. J., Schaefer, D. J., *Anal. Lett.*, **10**(4), 309–331 (1977).

Fourier transform ion cyclotron double resonance, Comisarow, M. B., Grassi, V., Parisod, G., *Chem. Phys. Lett.*, **57**(3), 413–416 (1978).

1,1-DICHLOROETHANE $C_2H_4Cl_2$ (98)

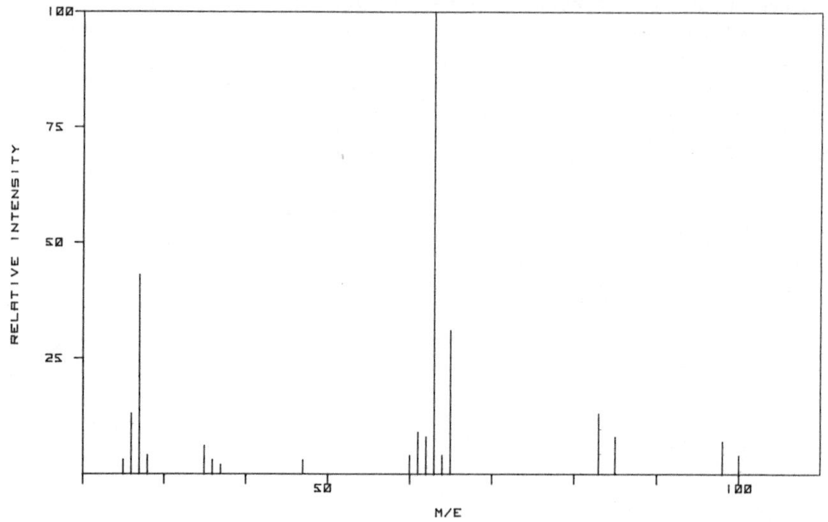

Spectral Data

Mass	Abundance	Mass	Abundance	Mass	Abundance
25	3.1	37	2.4	64	4.2
26	13.2	47	2.9	65	31.4
27	42.8	60	3.5	83	12.9
28	3.7	61	8.9	85	8.3
35	5.8	62	8.4	98	6.8
36	2.9	63	100.0	100	3.6

Volatile
CAS Name: Ethane, 1,1-dichloro-
Synonyms
- Chlorinated hydrochloric ether
- Ethylidene chloride
- Ethylidene dichloride
- NCI-C04535

CAS Registry No.: 75-34-3
ROTECS Ref.: KI01750
Merck Index Ref.: 3750

Major Ions: 63, 27, 65, 26, 83, 61, 62, 85
EPA Ions: 63, 65, 83, 85, 98, 100

Selected Bibliography

Negative ion mass spectra of chlorine-containing molecules, Ito, A., Matsumoto, K., Takeuchi, T., *Org. Mass Spectrom.*, **6**(9), 1045–1049 (1972).

Energy partitioning by mass spectrometry. Chloroalkanes and chloroalkenes, Kim, K. C., Beynon, J. H., Cooks, R. G., *J. Chem. Phys.*, **61**(4), 1305–1314 (1974).

A computer algorithm for qualitative identification of mass spectral data acquired in trace level analysis of environmental samples, Davidson, W. C., Smith, M. J., Schaefer, D. J., *Anal. Lett.*, **10**(4), 309–331 (1977).

1,2-DICHLOROETHANE $C_2H_4Cl_2$ (98)

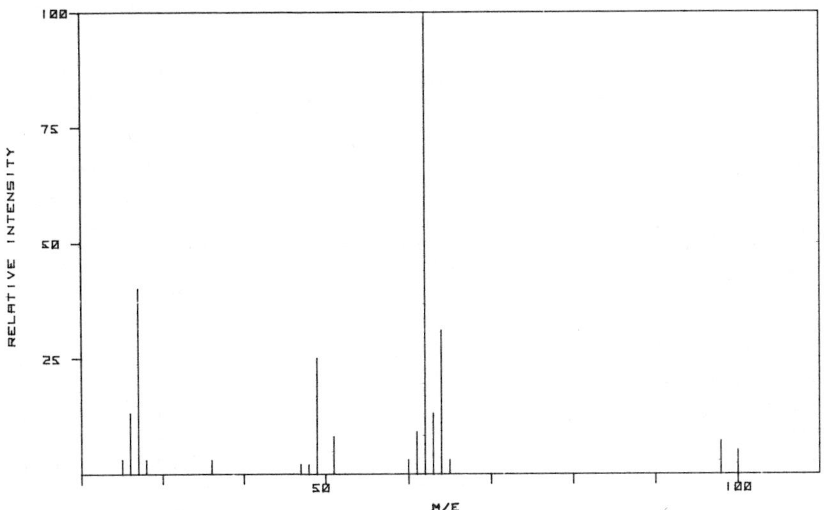

Spectral Data

Mass	Abundance	Mass	Abundance	Mass	Abundance
25	3.0	47	2.1	62	100.0
26	13.1	48	2.2	63	12.5
27	39.8	49	24.9	64	31.1
28	3.3	51	7.8	65	3.2
35	4.2	60	3.2	98	7.3
36	2.6	61	8.8	100	4.5

Volatile

CAS Name: Ethane, 1,2-dichloro-

Synonyms

1,2-Bichloroethane	Destruxol Borer-sol	Di-chlor-mulsion
Borer-sol	Dichloremulsion	*sym*-Dichloroethane
Brocide	1,2-Dichlorethane	

continued overleaf

1,2-DICHLOROETHANE–*continued*

Synonyms–*continued*

α,β-Dichloroethane
Dichloroethylene
Dutch liquid
EDC
Ent 1,656
Ethane dichloride
Ethylene chloride
Ethylene dichloride
Freon 150
Glycol dichloride
NCI-C00511
Oil of the Dutch Chemists

CAS Registry No.: 107-06-2 (formerly 52399-93-6)
ROTECS Ref.: KI05250
Merck Index Ref.: 3733

Major Ions: 62, 27, 64, 49, 26, 63, 61, 51
EPA Ions: 62, 64, 98, 100

Selected Bibliography

Negative ion mass spectra of chlorine-containing molecules, Ito, A., Matsumoto, K., Takeuchi, T., *Org. Mass Spectrom.*, **6**(9), 1045–1049 (1972).

Energy partitioning by mass spectrometry. Chloroalkanes and chloroalkenes, Kim, K. C., Beynon, J. H., Cooks, R. G., *J. Chem. Phys.*, **61**(4), 1305–1314 (1974).

Direct aqueous injection gas chromatography–mass spectrometry for analysis of organohalides in water at concentrations below the parts per billion level, Fujii, T., *J. Chromatogr.*, **139**(2), 297–302 (1977).

Kinetic energies of fission ions and repelling states of the initial molecular ion, Sumin, L. V., Gur'ev, M. V., *Teor. Eksp. Khim.*, **13**(6), 814–818 (1977).

A comparison of the translational energies released in the first field-free region after electron impact and field ionization, Derrick, P. J., Gardiner, T. M., Loudon, A. G., *Adv. Mass Spectrom.*, **7A**, 77–82 (1978).

HEXACHLOROETHANE C_2Cl_6 (234)

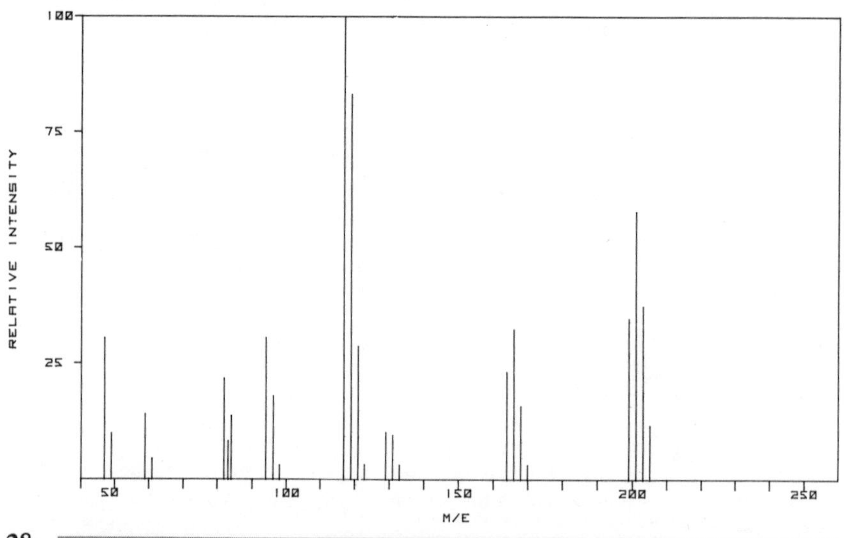

Spectral Data

Mass	Abundance	Mass	Abundance	Mass	Abundance
47	30.4	98	3.0	164	23.1
49	9.9	117	100.0	166	32.3
59	14.0	119	83.0	168	15.7
61	4.4	121	28.6	170	3.1
82	21.7	123	3.1	199	34.7
83	8.2	129	10.1	201	57.8
84	13.7	131	9.4	203	37.5
94	30.5	133	3.1	205	11.6
96	17.9				

Base–neutral extractable
CAS Name: Ethane, hexachloro-
Synonyms

Avlothane	Ethane hexachloride	Hexachloroethylene
Carbon hexachloride	Falkitol	Mottenhexe
Distokal	Fasciolin	NCI-C04604
Distopan	Hexachlorethane	Perchloroethane
Distopin	1,1,1,2,2,2-Hexachloroethane	Phenohep
Egitol		

CAS Registry No.: 67-72-1
ROTECS Ref.: KI40250
Merck Index Ref.: 4545

Major Ions: 117, 119, 201, 203, 199, 166, 94, 47
EPA Ions: 117, 201, 199

Selected Bibliography

Positive-ion mass spectra of some perhalogenated compounds, Contineanu, M. A., Grubel, L., *An. Univ. Bucuresti, Chim.*, **20**(2), 175–181 (1971).

Chlorine-35/chlorine-37 isotope effects in the fragmentations of metastable perchlorinated alkane and silane ions, Potzinger, P., Basu, S., *Ber. Bunsenges. Phys. Chem.*, **82**(4), 415–418 (1978).

1,1,2,2-TETRACHLOROETHANE $C_2H_2Cl_4$ (166)

Spectral Data

Mass	Abundance	Mass	Abundance	Mass	Abundance
25	3.1	61	11.9	98	6.0
26	8.1	62	4.6	130	2.6
35	8.4	63	3.8	131	8.7
36	2.6	83	100.0	132	2.9
37	2.8	85	66.3	133	8.4
45	2.0	87	9.8	135	2.6
47	5.4	95	15.6	166	7.7
48	4.1	96	11.1	168	9.9
49	2.5	97	10.8	170	4.1
60	14.1				

1,1,2,2-TETRACHLOROETHANE–*continued*

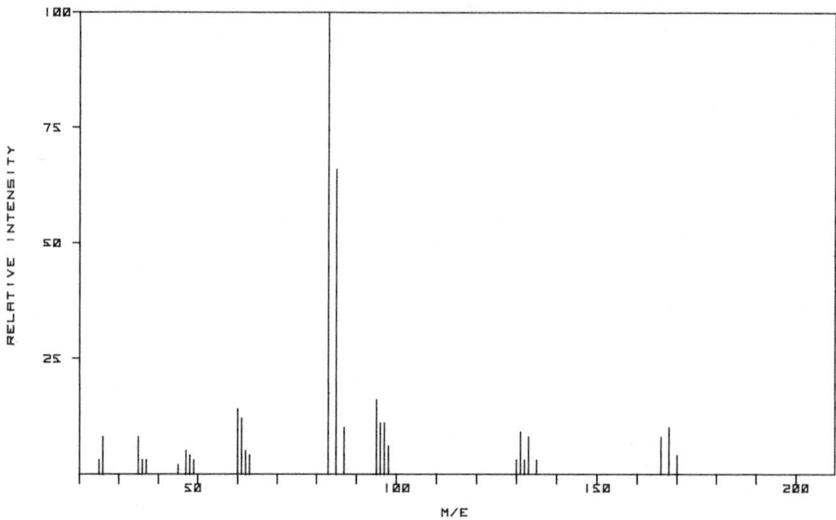

Volatile
CAS Name: Ethane, 1,1,2,2-Tetrachloro-
Synonyms
- Acetylene tetrachloride
- Bonoform
- Cellon
- NCI-C52459
- Tetrachloroethane
- *sym*-Tetrachloroethane

CAS Registry No.: 79-34-5
ROTECS Ref.: KI84000
Merck Index Ref.: 8906

Major Ions: 83, 85, 95, 60, 61, 96, 168, 87
EPA Ions: 83, 85, 131, 133, 166, 168

Selected Bibliography

Energy partitioning by mass spectrometry. Chloroalkanes and chloroalkenes, Kim, K. C., Beynon, J. H., Cooks, R. G., *J. Chem. Phys.*, **61**(4), 1305–1314 (1974).

A comparison of negative and positive ion mass spectrometry, Large, R., Knof, H., *Org. Mass Spectrom.*, **11**(6), 582–598 (1976).

1,1,1-TRICHLOROETHANE $C_2H_3Cl_3$ (132)

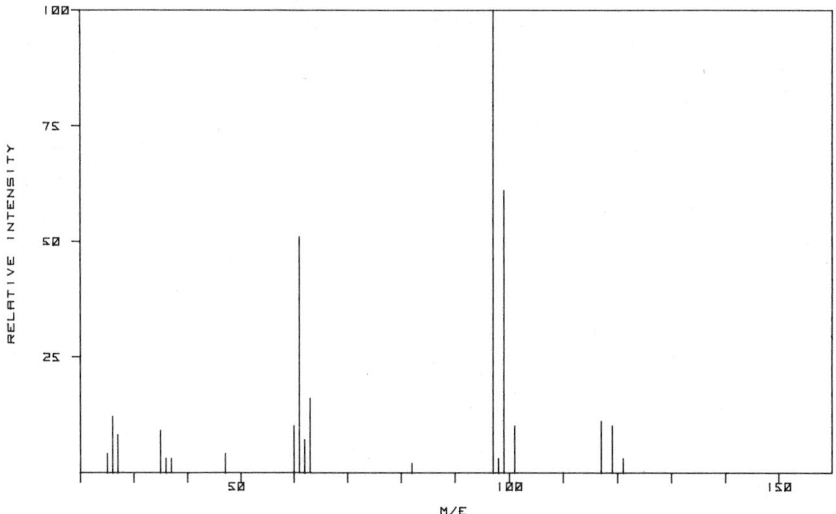

Spectral Data

Mass	Abundance	Mass	Abundance	Mass	Abundance
25	3.7	60	9.6	98	3.2
26	11.8	61	51.2	99	60.9
27	7.5	62	6.9	101	10.2
35	8.7	63	16.3	117	10.7
36	2.7	82	2.3	119	10.2
37	3.3	97	100.0	121	3.3
47	3.8				

Volatile
CAS Name: Ethane, 1,1,1-trichloro-
Synonyms

Aerothene TT	Chlorothene NU	Methyltrichloromethane
Chloroethene NU	Chlorothene VG	NCI-C04626
Chlorotene	Chlorten	α-T
Chlorothane NU	Inhibisol	Trichloroethane
Chlorothene	Methylchloroform	α-Trichloroethane

CAS Registry No.: 71-55-6
ROTECS Ref.: KJ29750
Merck Index Ref.: 9316

Major Ions: 97, 99, 61, 63, 26, 117, 119, 101
EPA Ions: 97, 99, 117, 119

1,1,1-TRICHLOROETHANE–*continued*

Selected Bibliography

Chlorinated hydrocarbons in the atmosphere. Analysis at the parts-per-trillion level by GC-MS (gas chromatography–mass spectrometry), Tyson, B. J., *Anal. Lett.*, 8(11), 807–813 (1975).

Direct aqueous injection gas chromatography–mass spectrometry for analysis of organohalides in water at concentrations below the parts per billion level, Fujii, T., *J. Chromatogr.*, **139**(2), 297–302 (1977).

1,1,2-TRICHLOROETHANE $C_2H_3Cl_3$ (132)

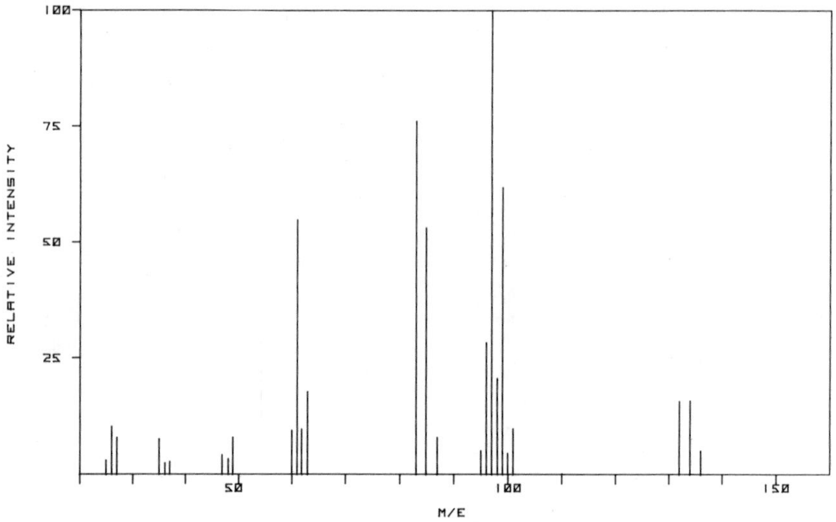

Spectral Data

Mass	Abundance	Mass	Abundance	Mass	Abundance
25	2.9	60	9.3	97	100.0
26	10.2	61	54.7	98	20.6
27	7.8	62	9.6	99	61.7
35	7.5	63	17.7	100	4.4
36	2.3	83	76.0	101	9.8
37	2.6	85	53.0	132	15.7
47	4.1	87	7.8	134	15.8
48	3.2	95	5.0	136	4.9
49	7.8	96	28.3		

Volatile
CAS Name: Ethane, 1,1,2-trichloro-
Synonyms

Ethane trichloride	β-Trichloroethane
NCI-C04579	1,2,2-Trichloroethane
β-T	Vinyl trichloride
1,1,2-Trichlorethane	

CAS Registry No.: 79-00-5
ROTECS Ref.: KJ31500
Merck Index Ref.: 9317

Major Ions: 97, 83, 99, 61, 85, 96, 98, 63
EPA Ions: 83, 85, 97, 99, 132, 134

Selected Bibliography

Energy partitioning by mass spectrometry. Chloroalkanes and chloroalkenes, Kim, K. C., Beynon, J. H., Cooks, R. G., *J. Chem. Phys.*, **61**(4), 1305–1314 (1974).

CHLOROALKYL ETHERS

Bis(2-chloroethyl) ether, bis(chloromethyl) ether, and 2-chloroethyl vinyl ether are included in this category. Bis(2-chloroisopropyl) ether is listed under "haloethers."

Bis(2-chloroethyl) ether has been used as an insecticidal soil fumigant, particularly in greenhouses.

2-Chloroethyl vinyl ether is used in the manufacture of anesthetics, cellulose ethers, and sedatives.

BIS(2-CHLOROETHYL) ETHER $C_4H_8Cl_2O$ (142)

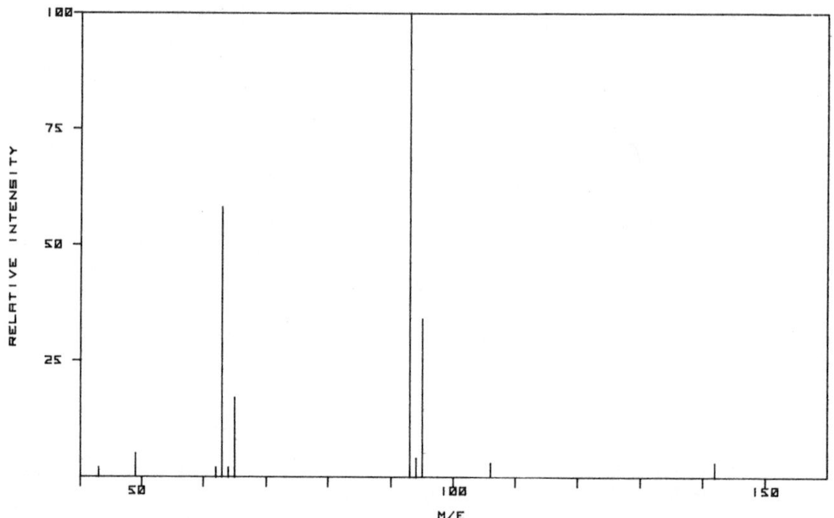

Spectral Data

Mass	Abundance	Mass	Abundance	Mass	Abundance
43	2.2	64	2.0	95	34.4
49	5.4	65	16.6	106	3.0
62	2.4	93	100.0	142	2.9
63	57.8	94	3.8		

Base–neutral extractable
CAS Name: Ether, bis(2-chloroethyl)
Synonyms

Bis(β-chloroethyl) ether
Bis(chloro-2-ethyl) oxide
Chlorex
1-Chloro-2-(β-chloroethoxy)ethane
Chloroethyl ether
2-Chloroethyl ether
DCEE
β,β-Dichlorodiethyl ether
Dichloroether
Dichloroethyl ether
Di(2-chloroethyl) ether
Di(β-chloroethyl) ether
β,β'-Dichloroethyl ether
sym-Dichloroethyl ether
2,2′-Dichloroethyl ether
Dichloroethyl oxide
Ent 4,504
1,1′-Oxybis(2-chloro)ethane

CAS Registry No.: 111-44-4
ROTECS Ref.: KN08750
Merck Index Ref.: 3040

Major Ions: 93, 63, 95, 65, 49, 94, 106, 142
EPA Ions: 93, 63, 95

BIS(CHLOROMETHYL) ETHER — C_2H_4ClO (114)

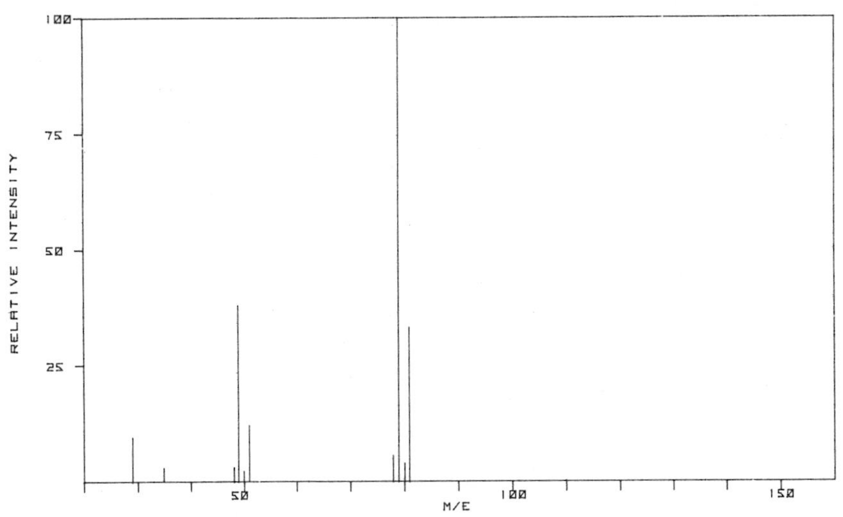

Spectral Data

Mass	Abundance	Mass	Abundance	Mass	Abundance
29	9.5	50	2.1	79	100.0
35	2.8	51	11.9	80	3.8
48	2.9	78	5.5	81	33.1
49	37.9				

BIS(CHLOROMETHYL)ETHER–*continued*

Volatile
CAS Name: Ether, bis(chloromethyl)
Synonyms

Bis-CME
BCME
Chloro(chloromethoxy)methane
Chloromethyl ether
α,α′-Dichlorodimethyl ether
sym-Dichloro-dimethyl ether
sym-Dichloromethyl ether
Dimethyl-1,1-dichloroether
Monochloromethyl ether
Oxybis(chloromethane)

CAS Registry No.: 542-88-1
ROTECS Ref.: KN15750
Merck Index Ref.: 3046

Major Ions: 79, 49, 81, 51, 29, 78, 80, 48
EPA Ions: 45, 49, 51

Selected Bibliography

Analysis of a noncrosslinked, water-soluble anion-exchange resin for the possible presence of parts per billion level of bis(chloromethyl) ether, Tou, J. C., Westover, L. B., Sonnabend, L. F., *J. Am. Ind. Hyg. Assoc.*, **36**(5), 374–378 (1975).

2-CHLOROETHYL VINYL ETHER C_4H_7ClO (106)

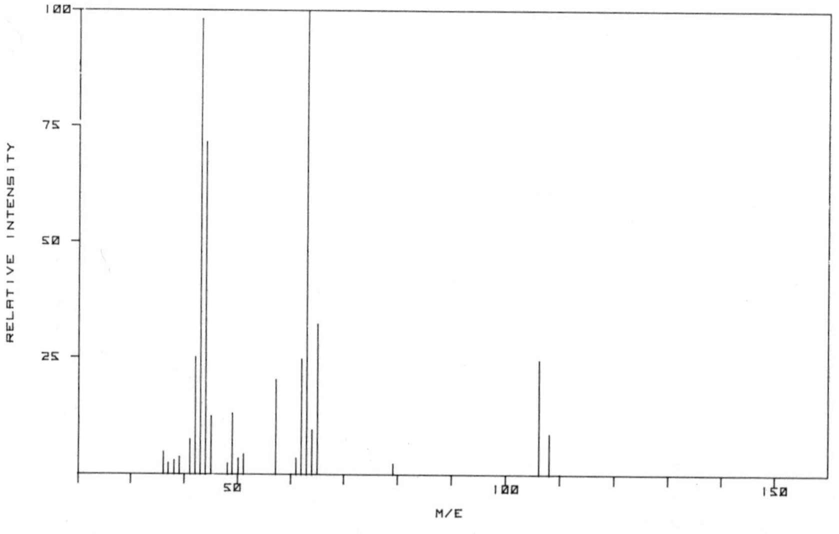

Spectral Data

Mass	Abundance	Mass	Abundance	Mass	Abundance
36	4.7	45	12.3	62	24.7
37	2.3	48	2.3	63	100.0
38	2.8	49	13.0	64	9.5
39	3.6	50	3.3	65	32.3
41	7.4	51	4.2	79	2.3
42	25.0	57	20.2	106	24.5
43	98.1	61	3.4	108	8.6
44	71.5				

Volatile
CAS Name: Ether, 2-chloroethyl vinyl
Synonyms
- β-Chloroethyl vinyl ether
- Vinyl β-chloroethyl ether
- Vinyl 2-chloroethyl ether
- 2-Vinyloxyethyl chloride

CAS Registry No.: 110-75-8
ROTECS Ref.: KN63000
Merck Index Ref.: 2119

Major Ions: 63, 43, 44, 65, 42, 62, 106, 57
EPA Ions: 63, 65, 106

Selected Bibliography

Metastable ion characteristics. XXXV. Structure and formation of stable $C_2H_4O^{+\cdot}$ ions. Van de Sande, C. C., McLafferty, F. W., *J. Am. Chem. Soc.*, 97(16), 4613–4616 (1975).

CHLORINATED NAPHTHALENE

2-Chloronaphthalene is the specific compound included in this category.

This compound was formerly used as an insecticide. 1-Chloronaphthalene (not listed as a priority pollutant) has been used as a solvent for DDT.

2-CHLORONAPHTHALENE $C_{10}H_7Cl$ (162)

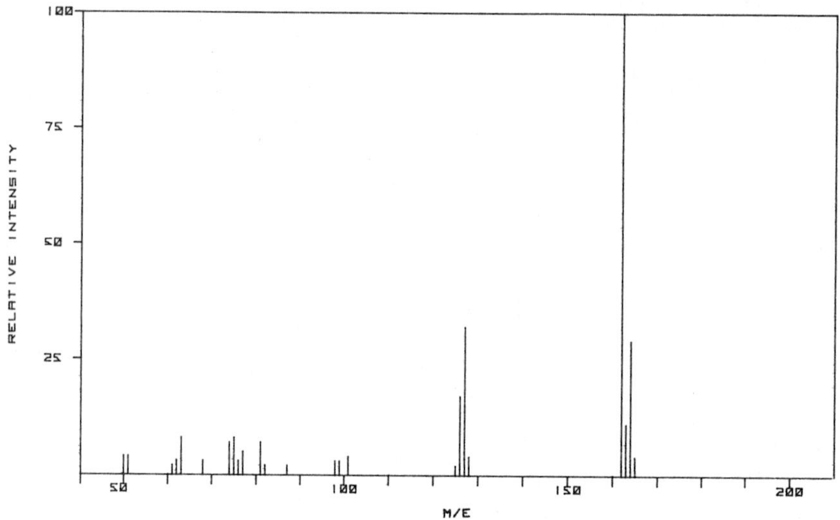

Spectral Data

Mass	Abundance	Mass	Abundance	Mass	Abundance
50	4.3	76	3.0	125	2.3
51	4.0	77	5.1	126	17.3
61	2.3	81	6.7	127	32.1
62	3.4	82	2.3	128	3.6
63	8.3	87	2.4	162	100.0
68	2.7	98	2.5	163	10.9
74	6.5	99	2.5	164	29.0
75	7.6	101	3.6	165	3.5

CHLORINATED NAPHTHALENE

Base–neutral extractable
CAS Name: Naphthalene, 2-chloro-
Synonym
β-Chloronaphthalene

CAS Registry No.: 91-58-7
ROTECS Ref.: QJ22750
Merck Index Ref.: 2127

Major Ions: 162, 127, 164, 126, 163, 63, 75, 81
EPA Ions: 162, 164, 127

Selected Bibliography

Ion kinetic energy and mass spectra of the isomeric mono- and dichloronaphthalenes, Safe, S., Hutzinger, O., Cook, M., *J. Chem. Soc., Chem. Commun.*, (5), 260–261, (1972).

CHLORINATED PHENOLS
(other than those listed elsewhere)

2,4,6-Trichlorophenol and *p*-chloro-*m*-cresol are included in this category. 2-Chlorophenol, 2,4-dichlorophenol, and pentachlorophenol (*q.v.*) are listed elsewhere.

2,4,6-Trichlorophenol is used as a fungicide and bactericide, especially for the control of fungal slimes. This compound is almost insoluble, but the sodium salt is water soluble.

p-Chloro-*m*-cresol is an antiseptic and disinfectant. It is registered for use as a fungicide in adhesives.

2,4,6-TRICHLOROPHENOL $C_6H_3Cl_3O$ (196)

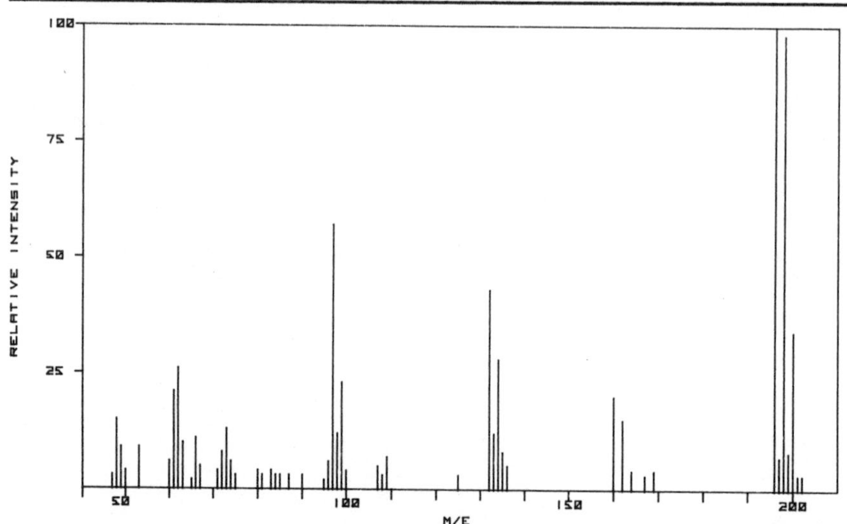

Spectral Data

Mass	Abundance	Mass	Abundance	Mass	Abundance
47	3.4	61	21.2	71	3.7
48	14.7	62	25.7	72	7.9
49	9.0	63	9.9	73	13.3
50	3.7	65	2.3	74	5.9
53	8.5	66	10.7	75	2.5
60	6.2	67	4.5	80	4.0

Spectral Data–*continued*

Mass	Abundance	Mass	Abundance	Mass	Abundance
81	2.8	100	3.7	162	15.0
83	4.0	107	5.4	164	3.7
84	2.8	108	3.4	167	3.4
85	3.1	109	7.1	169	3.7
87	3.4	125	2.5	196	100.0
90	2.5	132	42.7	197	7.3
95	2.3	133	12.4	198	97.5
96	5.9	134	28.0	199	7.6
97	56.8	135	7.9	200	33.9
98	12.1	136	4.8	201	2.8
99	22.9	160	19.5	202	2.8

Acid extractable
CAS Name: Phenol, 2,4,6-trichloro-
Synonyms
- Dowicide 2
- Dowicide 2S (sodium salt)
- NCI-C02904
- Omal
- Phenachlor

CAS Registry No.: 88-06-2
ROTECS Ref.: SN15750
Merck Index Ref.: 9323

Major Ions: 196, 198, 97, 132, 200, 134, 62, 99
EPA Ions: 196, 198, 200

Selected Bibliography

Gas–liquid chromatography as an analytical tool in the production of pentachlorophenol and hexachlorophenol, Svec, P., Zbirovsky, M., *Sp. Vys. Sk. Chem-Technol. Praze, Org. Chem. Technol.*, **C21**, 39–43 (1974).

p-CHLORO-*m*-CRESOL — C_7H_7ClO (142)

Spectral Data

Mass	Abundance	Mass	Abundance	Mass	Abundance
50	8.5	73	3.9	107	100.0
51	17.4	74	3.1	108	8.3
52	3.6	75	4.1	113	3.4
53	7.7	77	43.7	141	7.8
61	2.9	78	12.1	142	76.4
62	4.9	79	11.8	143	9.0
63	7.3	89	2.6	144	24.8
71	2.2	105	6.3	145	2.2

p-CHLORO-*m*-CRESOL–*continued*

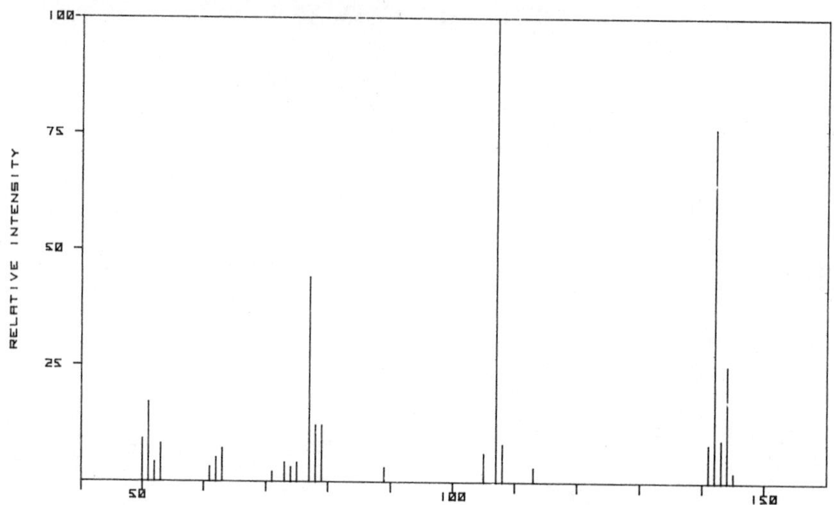

Acid extractable
CAS Name: *m*-Cresol, 4-chloro-
Synonyms

- Aptal
- Baktol
- Baktolan
- Candaseptic
- *p*-Chlor-*m*-cresol
- Chlorocresol
- *p*-Chlorocresol
- 4-Chloro-*m*-cresol
- 6-Chloro-*m*-cresol
- 2-Chloro-hydroxytoluene
- 2-Chloro-5-hydroxytoluene
- 6-Chloro-3-hydroxytoluene
- 4-Chloro-3-methylphenol
- 4-Chloro-5-methylphenol
- 3-Methyl-4-chlorophenol
- Ottafact
- Parmetol
- Parol
- PCMC
- Peritonan
- Preventol CMK
- Raschit
- Raschit K
- Rasen-Anicon

CAS Registry No.: 59-50-7
ROTECS Ref.: GO71000
Merck Index Ref.: 2108

Major Ions: 107, 142, 77, 144, 51, 78, 79, 143
EPA Ions: 142, 107, 144

CHLOROFORM

Chloroform is a solvent for many substances, including fats, oils, rubber, alkaloids, waxes, gutta-percha, and resins. It has been used as an anesthetic, an insect fumigant, a cleansing agent, and in fire extinguishers (to depress the freezing point of carbon tetrachloride). This compound is usually the major trihalomethane found in chlorinated water.

CHLOROFORM **$CHCl_3$ (118)**

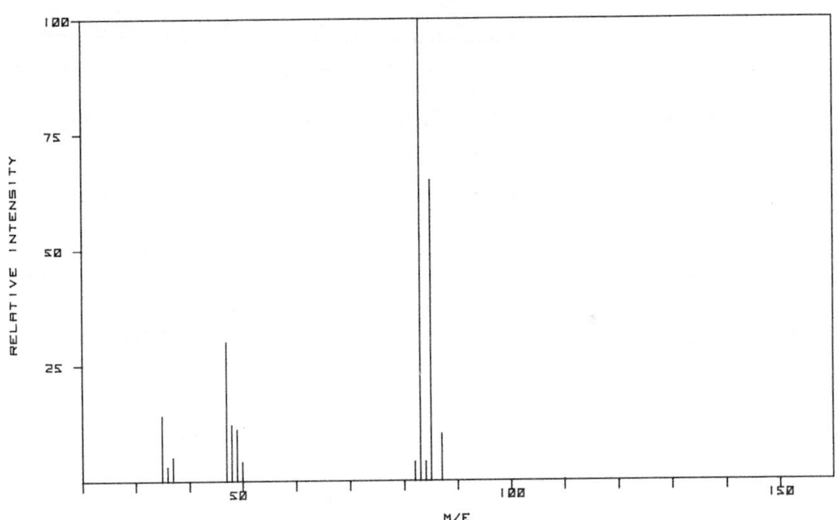

Spectral Data

Mass	Abundance	Mass	Abundance	Mass	Abundance
35	14.4	48	11.9	83	100.0
36	2.5	49	10.5	84	3.8
37	4.8	50	3.9	85	64.7
47	29.7	82	4.4	87	9.5

CHLOROFORM–*continued*

Volatile
CAS Name: Chloroform
Synonyms

Formyl trichloride	NCI-C02686
Freon 20	R 20
Methane trichloride	Trichloroform
Methylene trichloride	Trichloromethane

CAS Registry No.: 67-66-3 (formerly 8013-54-5)
ROTECS Ref.: FS91000
Merck Index Ref.: 2120

Major Ions: 83, 85, 47, 35, 48, 49, 87, 37
EPA Ions: 83, 85

Selected Bibliography

The determination of traces of organohalogen compounds in aqueous solution by direct injection gas chromatography–mass spectrometry and single ion detection, Fujii, T., *Anal. Chim. Acta*, **92**(1), 117–122 (1977).

Direct aqueous injection gas chromatography–mass spectrometry for analysis of organohalides in water at concentrations below the parts per billion level, Fujii, T., *J. Chromatogr.*, **139**(2), 297–302 (1977).

Volatile substances in blood serum: profile analysis and quantitative determination, Leibich, H. M., Woell, J., *J. Chromatogr.*, **142**, 505–516 (1977).

Head space mass spectrometric analysis for volatiles in biological specimens, Urich, R. W., Bowerman, D. L., Wittenberg, P. H., Pierce, A. F., Schisler, D. K., Levisky, J. A., Pflug, J. L., *J. Anal. Toxicol.*, **1**(5), 195–199 (1977).

Concentration and analysis of trace impurities in styrene monomer, Zlatkis, A., Anderson, J. W., Holzer, G., *J. Chromatogr.*, **142**, 127–129 (1977).

2-CHLOROPHENOL

See, also, "chlorinated phenols."

2-CHLOROPHENOL C_6H_5ClO (128)

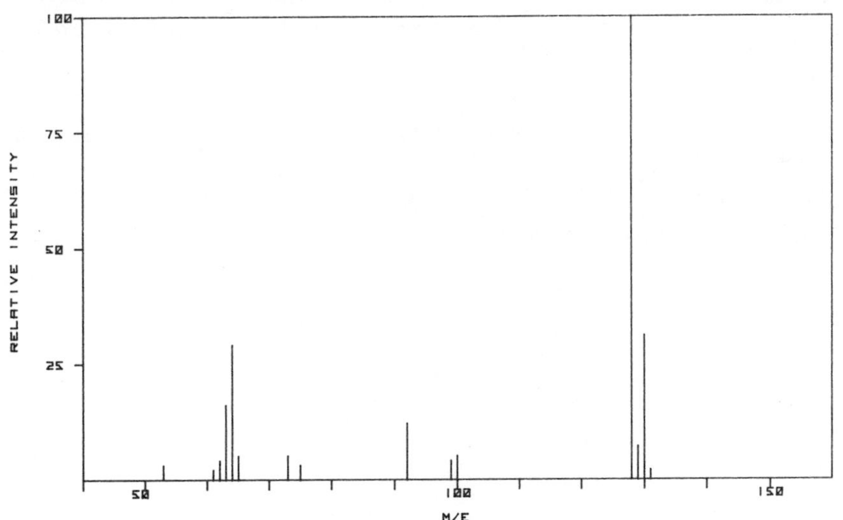

Spectral Data

Mass	Abundance	Mass	Abundance	Mass	Abundance
53	2.6	65	4.8	100	4.7
61	2.3	73	5.0	128	100.0
62	3.7	75	2.9	129	6.7
63	16.1	92	12.3	130	31.2
64	29.3	99	3.9	131	2.2

Acid extractable
CAS Name: Phenol, *o*-chloro-
Synonyms
- *o*-Chlorophenol
- 2-Chlorophenol
- 2-Hydroxychlorobenzene

2-CHLOROPHENOL–*continued*

CAS Registry No.: 95-57-8
ROTECS Ref.: SK26250
Merck Index Ref.: 2134

Major Ions: 128, 130, 64, 63, 92, 129, 73, 65
EPA Ions: 128, 64, 130

Selected Bibliography

Interpretation of mass spectra by simple LCAO-MO calculation. II. Derivatives of benzene and chlorobenzene, Tajima, S., Niwa, Y., Wasada, N., Tsuchiya, T., *Bull. Chem. Soc. Jap.*, **45**(4), 1250–1251 (1972).

Doubly charged ion mass spectra. IV. Chlorobenzene derivatives, Sakurai, H., Tatematsu, A., Nakata, H., *Shitsuryo Bunseki*, **23**(4), 291–298 (1975).

Temperature dependence of mass spectra on conformational transformation of ortho-substituted phenols in the gas phase, Bogolyubov, G. M., Gal'perin, Ya. V., Petrov, A. A., *Zh. Obshch. Khim.*, **46**(2), 336–340 (1976).

Collisionally-induced decompositions, Goren, A., Munson, B., Shimizu, Y., *Int. J. Mass Spectrom. Ion Phys.*, **21**(1–2), 73–80 (1976).

DDT AND METABOLITES

4,4′-DDD, 4,4′-DDE, and 4,4′-DDT are included in this category.

4,4′-DDT is the major component (70%–80%) of the insecticide DDT. Its use is now severely restricted. Environmental degradation and metabolism of 4,4′-DDT yields, 4,4′-DDD and 4,4′-DDE. 4,4′-DDD is also the principal constituent of the insecticide TDE.

4,4′-DDD should not be confused with 2,2′-dihydroxy-6,6′-dinaphthyl disulfide (also known as DDD) used for the estimation of protein-bound sulfhydryl groups.

4,4′-DDD $C_{14}H_{10}Cl_4$ (318)

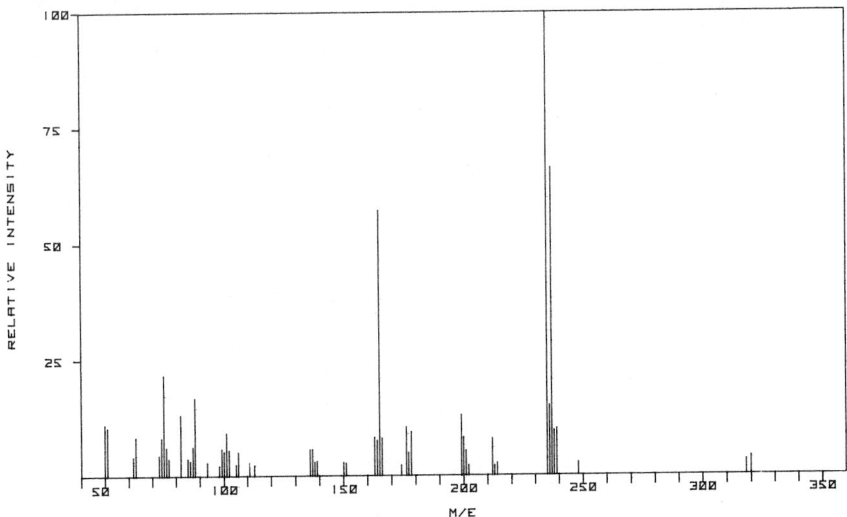

Spectral Data

Mass	Abundance	Mass	Abundance	Mass	Abundance
50	9.2	74	6.8	85	3.9
51	9.2	75	21.2	86	2.7
62	3.6	76	4.6	87	5.3
63	6.9	77	2.8	88	12.3
67	3.4	82	10.2	89	3.1
73	2.6	83	5.8	93	2.3

4,4′-DDD–*continued*

Spectral Data–*continued*

Mass	Abundance	Mass	Abundance	Mass	Abundance
98	2.0	151	2.9	201	5.5
99	5.5	163	7.6	202	2.2
100	4.8	164	7.8	212	5.8
101	9.6	165	58.4	214	2.1
102	5.6	166	8.3	235	100.0
106	4.1	174	2.3	236	12.2
111	2.8	175	2.4	237	57.0
136	6.0	176	9.2	238	9.1
137	5.5	177	4.3	239	10.0
138	3.0	178	7.5	318	2.6
139	3.3	199	13.5	320	3.5
150	2.2	200	8.0		

Pesticide
CAS Name: Ethane, 1,1-dichloro-2,2-bis(*p*-chlorophenyl)-
Synonyms

1,1-Bis(*p*-chlorophenyl)-2,2-dichloroethane
1,1-Bis(4-chlorophenyl)-2,2-dichloroethane
2,2-Bis(*p*-chlorophenyl)-1,1-dichloroethane
2,2-Bis(4-chlorophenyl)-1,1-dichloroethane
DDD
p,p′-DDD
1,1-Dichloro-2,2-bis(*p*-chlorophenyl)ethane
1,1-Dichloro-2,2-bis(parachlorophenyl)ethane
1,1-Dichloro-2,2-di(4-chlorophenyl)ethane
Dichlorophenyl dichloroethane
p,p′-Dichlorodiphenyl dichloroethane
p,p′-Dichlorodiphenyl-2,2-dichloroethylene
Dilene
Ent 4,225
ME-1,700
NCI-C00475
Rhothane
Rhothane D-3
Rothane
TDE
p,p′-TDE
Tetrachlorodiphenylethane

CAS Registry No.: 72-54-8
ROTECS Ref.: KI07000
Merck Index Ref.: 3034

Major Ions: 235, 165, 237, 75, 199, 88, 236, 82
EPA Ions: 235, 165, 237

Selected Bibliography

Coupling of a liquid chromatograph to a mass spectrometer, Lovins, R. E., Ellis, S. R., Tolbert, G. D., McKinney, C. R., *Adv. Mass Spectrom.*, **6**, 457–462 (1974).

Mass-spectrometric identification and quantitation of chlorinated hydrocarbons in fish, Schaefer, R. G., *Chem.-Ztg.*, **98**(5), 241–247 (1974).

Application of coupled gas chromatography–mass spectrometry in methods for the study and determination of pesticide residues and organic micropollutants in environmental and food materials, Mestres, R., Chevallier, C., Espinoza, C., Cornet, R., *Ann. Falsif. Exper. Chim.*, **70**(751), 177–188 (1977).

4,4′-DDE $C_{14}H_8Cl_4$ (316)

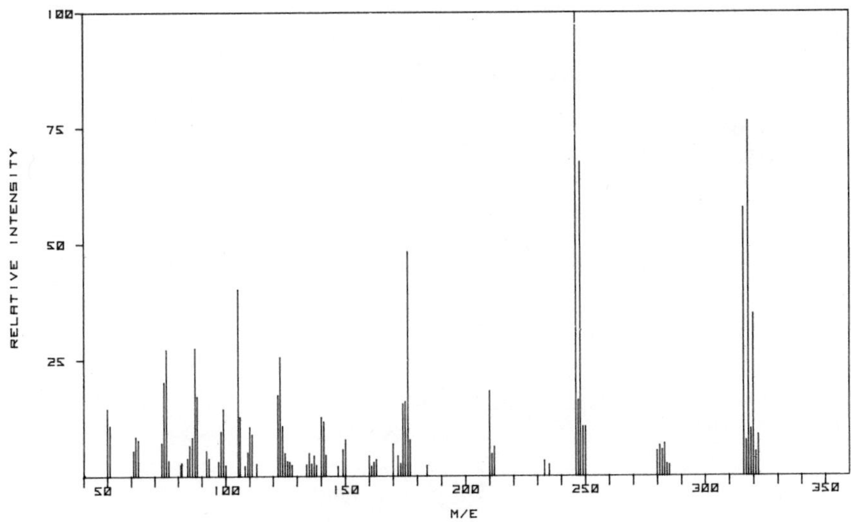

Spectral Data

Mass	Abundance	Mass	Abundance	Mass	Abundance
50	13.2	124	8.8	211	4.6
61	3.3	125	4.9	212	6.0
62	3.3	126	2.7	233	2.8
63	2.4	134	2.4	246	100.0
74	14.9	135	2.8	247	14.0
75	23.8	137	2.8	248	72.9
76	2.6	140	11.8	249	9.8
86	3.8	141	8.9	250	10.2
87	22.3	142	3.1	280	3.4
88	14.4	149	2.3	282	5.8
92	4.6	150	8.0	283	5.5
93	3.7	160	4.2	284	2.3
98	7.2	163	3.8	285	2.4
99	12.7	170	4.0	316	57.1
105	40.4	172	5.4	317	8.3
106	11.0	174	14.8	318	76.8
110	8.2	175	12.4	319	9.5
111	6.0	176	57.0	320	35.0
122	15.5	177	6.2	321	4.8
123	21.5	210	20.9	322	8.1

4,4′-DDE–*continued*

Pesticide
CAS Name: Ethylene, 1,1-dichloro-2,2-bis(*p*-chlorophenyl)-
Synonyms

- 2,2-Bis(*p*-chlorophenyl)-1,1-dichloroethylene
- DDE
- *p,p*′-DDE
- DDT dehydrochloride
- DDX
- 1,1-Dichloro-2,2-bis(*p*-chlorophenyl)ethylene
- *p,p*′-Dichlorodiphenyldichloroethylene
- 1,1′-Dichloroethenylidene-bis(4-chlorobenzene)
- NCI-C00555

CAS Registry No.: 72-55-9
ROTECS Ref.: KV94500

Major Ions: 246, 318, 248, 316, 176, 105, 320, 75
EPA Ions: 246, 248, 176

Selected Bibliography

Ion kinetic energy spectra of some chlorinated insecticides, Safe, S., Hutzinger, O., Jamieson, W. D., Cook, M., *Org. Mass Spectrom.*, 7(2), 217–224 (1973).

Mass-spectrometric identification and quantitation of chlorinated hydrocarbons in fish, Schaefer, R. G., *Chem.-Ztg.*, **98**(5), 241–247, (1974).

Application of coupled gas chromatography–mass spectrometry in methods for the study and determination of pesticide residues and organic micropollutants in environmental and food materials, Mestres, R., Chevallier, C., Espinoza, C., Cornet, R., *Ann. Falsif. Exper. Chim.*, **70**(751), 177–188 (1977).

4,4′-DDT

$C_{14}H_9Cl_5$ (352)

Spectral Data

Mass	Abundance	Mass	Abundance	Mass	Abundance
50	8.8	86	3.2	106	4.6
51	6.9	87	6.7	110	2.4
62	3.4	88	9.9	111	3.4
63	5.8	89	2.2	113	2.3
73	3.8	93	2.1	123	6.0
74	8.4	98	2.6	124	3.3
75	17.6	99	6.6	135	3.2
76	2.9	100	3.7	136	8.5
82	8.3	101	4.2	137	2.6
85	2.7	105	6.3	138	3.5

Spectral Data—*continued*

Mass	Abundance	Mass	Abundance	Mass	Abundance
139	2.0	176	13.3	235	100.0
150	2.4	177	3.0	236	12.5
151	2.2	199	11.9	237	56.6
163	7.7	200	7.3	238	8.9
164	7.4	201	5.4	239	9.8
165	53.8	202	2.3	246	5.3
166	8.0	210	2.0	247	2.3
171	2.2	212	11.0	248	4.0
173	2.1	213	2.4	282	3.3
174	2.8	214	3.9	284	3.2
175	2.9				

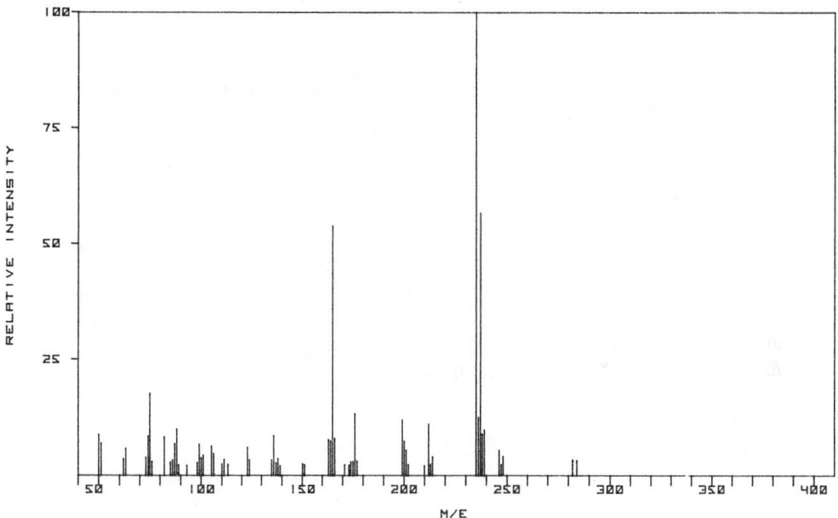

Pesticide
CAS Name: Ethane, 1,1,1-trichloro-2,2-bis(*p*-chlorophenyl)-
Synonyms

- Aavero-extra
- Agritan
- Anofex
- Arkotine
- Azotox
- Azotox M-33
- 2,2-Bis(*p*-chlorophenyl)-1,1,1-trichloroethane
- α,α-Bis(*p*-chlorophenyl)-β,β,β-trichloroethane
- Bosan supra
- Bovidermol
- Chlorophenotane
- Chlorphenothane
- Chlorphenotoxum
- Clofenotane
- DDT
- *p,p'*-DDT
- Deoval
- Detox
- Detoxan

continued overleaf

4,4′-DDT–*continued*

Synonyms–*continued*

Dibovin
4,4′-Dichlorodiphenyltrichloroethane
p,p′-Dichlorophenyltrichloroethane
Dicophane
Didigam
Didimac
Dinocide
Dodat
Dykol
Ent 1,506
Estonate
Genitox
Gesafid
Gesapon
Gesarex
Gesarol
Guesarol
Gyron
Ixodex
Kopsol
Mutoxan
NCI-C00464
Neocid
Neocidol
Parachlorocidum
PEB1
Pentachlorin
Pentech
Penticidum
Ppzeidan
Rukseam
Santobane
1,1,1-Trichloro-2,2-bis(4-chlorophenyl)ethane
1,1,1-Trichloro-2,2-di(4-chlorophenyl)ethane
Zeidane
Zerdane

CAS Registry No.: 50-29-3
ROTECS Ref.: KJ33250
Merck Index Ref.: 2822

Major Ions: 235, 237, 165, 75, 176, 236, 199, 212
EPA Ions: 235, 237, 165

Selected Bibliography

Mass-spectrometric identification and quantitation of chlorinated hydrocarbons in fish, Schaefer, R. G., *Chem.-Ztg.*, **98**(5), 241–247 (1974).

Application of coupled gas chromatography–mass spectrometry in methods for the study and determination of pesticide residue and organic micropollutants in environmental and food materials, Mestres, R., Chevallier, C., Espinoza, C., Cornet, R., *Ann. Falsif. Exper. Chim.*, **70**(751), 177–188 (1977).

Quantitative analysis of biologically active substances by mass fragmentography. Precision of the method, Eyem, J., *Adv. Mass Spectrom.*, **7B**, 1534–1539 (1978).

Some considerations on quantitative methodology and detection limits in organic mass spectrometry, Marshall, D. J., Petersen, B. A., Vouros, P., *Biomed. Mass Spectrom.*, **5**(3), 243–249 (1978).

Application of new mass spectrometry techniques for gas chromatography/mass spectrometry routine analysis in environmental chemistry, food control and related fields, Naegeli, P., Egli, H. P., *Adv. Mass Spectrom.*, **7B**, 1713–1720 (1978).

DICHLOROBENZENES

1,2-, 1,3-, and 1,4-dichlorobenzenes are included this category.

1,2-Dichlorobenzene is a solvent for many substances, including asphalts, gums, oils, resins, tars, and waxes. It is used for degreasing leather, metals, and wool. This compound is also used for removing sulfur from gas, in the manufacture of dyes, in metal polishes, and as a heat transfer medium. It is an insecticide effective against termites and locust borers.

1,4-Dichlorobenzene is used as a moth repellant. It is toxic to moth larvae and cockroaches.

1,2-DICHLOROBENZENE $C_6H_4Cl_2$ (146)

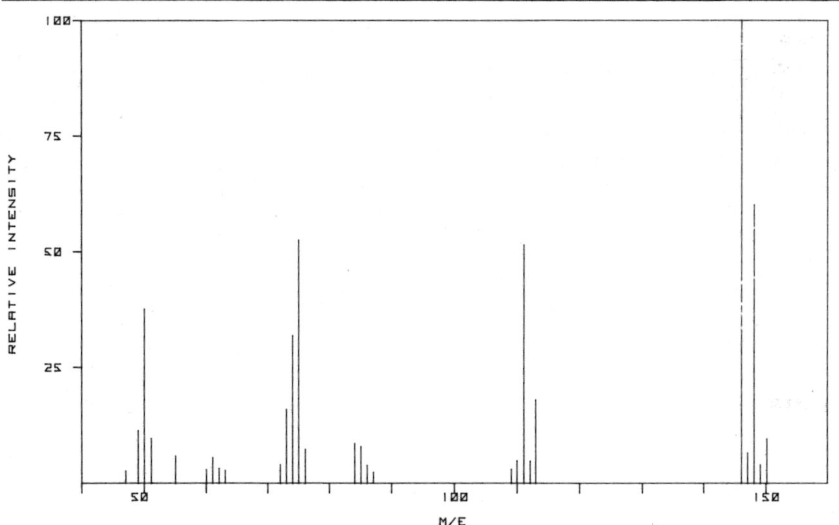

Spectral Data

Mass	Abundance	Mass	Abundance	Mass	Abundance
47	2.5	73	15.9	110	4.8
49	11.3	74	31.8	111	51.5
50	37.5	75	52.5	112	4.7
51	9.6	76	7.2	113	18.0
55	5.8	84	8.4	146	100.0
60	2.8	85	7.8	147	6.4
61	5.4	86	3.8	148	60.1
62	3.1	87	2.3	149	3.9
63	2.7	109	3.0	150	9.5
72	3.9				

1,2-DICHLOROBENZENE—*continued*

Base–neutral extractable
CAS Name: Benzene, *o*-dichloro-
Synonyms

Chloroben	*o*-Dichlorobenzene	ODB
Cloroben	Dilantin DB	ODCB
DCB	Dizene	Orthodichlorobenzol
o-Dichlor benzol	Dowtherm E	Special Termite Fluid
o-Dichlorbenzene	NCI-C54944	Termitkil

CAS Registry No.: 95-50-1
ROTECS Ref.: CZ45000
Merck Ref.: 3029

Major Ions: 146, 148, 75, 111, 50, 74, 113, 73
EPA Ions: 146, 148, 113

Selected Bibliography

Ion kinetic energy spectra of isomeric chlorobenzenes and polychlorinated biphenyls, Safe, S., Hutzinger, O., Jamieson, W. D., *Org. Mass Spectrom.*, 7(2), 169–176 (1973).

Doubly charged ion mass spectra. IV. Chlorobenzene derivatives, Sakurai, H., Tatematsu, A., Nakata, H., *Shitsuryo Bunski*, **23**(4), 291–298 (1975).

Fragmentation diagrams for the elucidation of decomposition reactions of organic compounds. II. Polymethyl- and chlorobenzenes, Schoenfeld, W., *Org. Mass Spectrom.*, **10**(6), 401–426 (1975).

1,3-DICHLOROBENZENE $C_6H_4Cl_2$ (146)

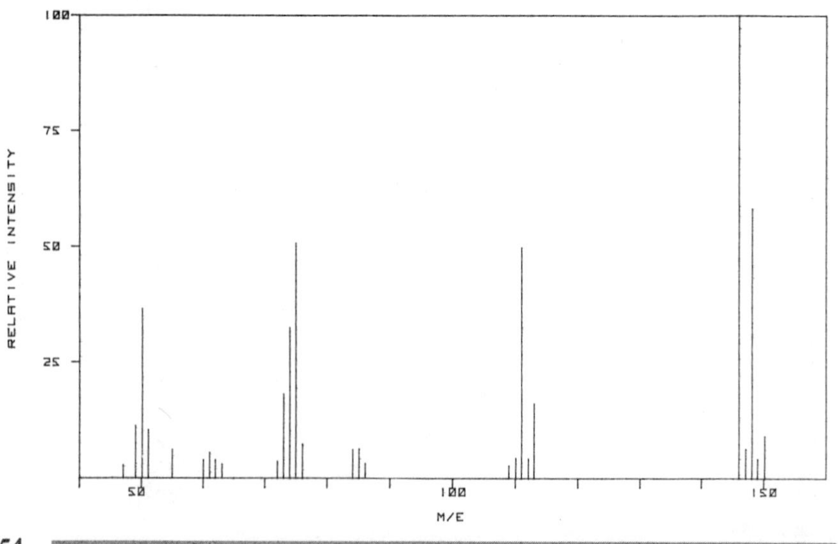

Spectral Data

Mass	Abundance	Mass	Abundance	Mass	Abundance
47	2.8	72	3.7	110	4.3
49	11.3	73	18.1	111	49.7
50	36.5	74	32.4	112	4.2
51	10.4	75	50.6	113	16.0
55	6.2	76	7.3	146	100.0
60	4.0	84	6.2	147	6.3
61	5.5	85	6.3	148	58.3
62	4.0	86	3.2	149	4.1
63	3.0	109	2.7	150	9.0

Base–neutral extractable
CAS Name: Benzene, *m*-dichloro-
Synonyms

m-Dichlorobenzene   *m*-Dichlorobenzol   *m*-Phenylene dichloride

CAS Registry No.: 541-73-1
Merck Ref.: 3028

Major Ions: 146, 148, 75, 111, 50, 74, 73, 113
EPA Ions: 146, 148, 113

Selected Bibliography

Ion kinetic energy spectra of isomeric chlorobenzenes and polychlorinated biphenyls, Safe, S., Hutzinger, O., Jamieson, W. D., *Org. Mass Spectrom.*, 7(2), 169–176 (1973).

Doubly charged ion mass spectra. IV. Chlorobenzene derivatives, Sakurai, H., Tatematsu, A., Nakata, H., *Shitsuryo Bunseki*, 23(4), 291–298 (1975).

1,4-DICHLOROBENZENE $C_6H_4Cl_2$ (146)

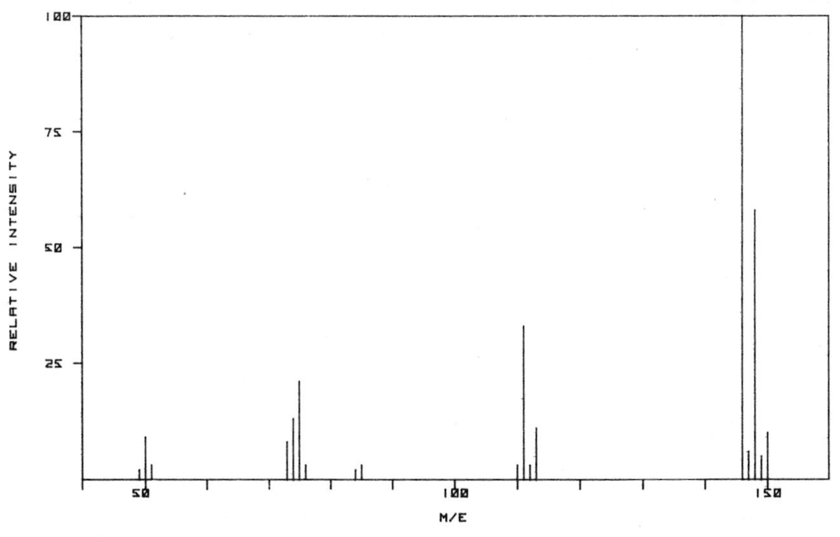

1,4-DICHLOROBENZENE—*continued*

Spectral Data

Mass	Abundance	Mass	Abundance	Mass	Abundance
49	2.4	76	2.5	113	11.0
50	9.4	84	2.4	146	100.0
51	2.9	85	2.9	147	6.0
73	7.5	110	2.7	148	58.2
74	13.0	111	33.4	149	4.5
75	20.8	112	2.6	150	10.1

Base–neutral extractable
CAS Name: Benzene, *p*-dichloro-
Synonyms

p-Chlorophenyl chloride
Di-chloricide
p-Dichlorobenzene
p-Dichlorobenzol
Evola
NCI-C54955
Paracide
Paradi
Paradichlorobenzol
Paradow
Paramoth
Paranuggets
Parazene
PDB
Persia-Perazol
Santochlor

CAS Registry No.: 106-46-7
ROTECS Ref.: CZ45500
Merck Ref.: 3030

Major Ions: 146, 148, 111, 75, 74, 113, 150, 50
EPA Ions: 146, 148, 113

Selected Bibliography

Effects of induction and resonance in the calculation of ionization potentials of substituted benzenes by perturbation molecular orbital theory, Johnstone, R. A., Mellon, F. A., *J. Chem. Soc., Faraday Trans. 2*, **69**(1), 36–42 (1973).

Ion kinetic energy spectra of isomeric chlorobenzenes and polychlorinated biphenyls, Safe, S., Hutzinger, O., Jamieson, W. D., *Org. Mass Spectrom.*, 7(2), 169–176 (1973).

Doubly charged ion mass spectra. IV. Chlorobenzene derivatives, Sakurai, H., Tatematsu, A., Nakata, H., *Shitsuryo Bunseki*, **23**(4), 291–298 (1975).

DICHLOROBENZIDINE

3,3′-Dichlorobenzidine is the specific compound included in this category. This compound has been used in the manufacture of azo dyes.

3,3′-DICHLOROBENZIDINE $C_{12}H_{10}Cl_2N_2$ (252)

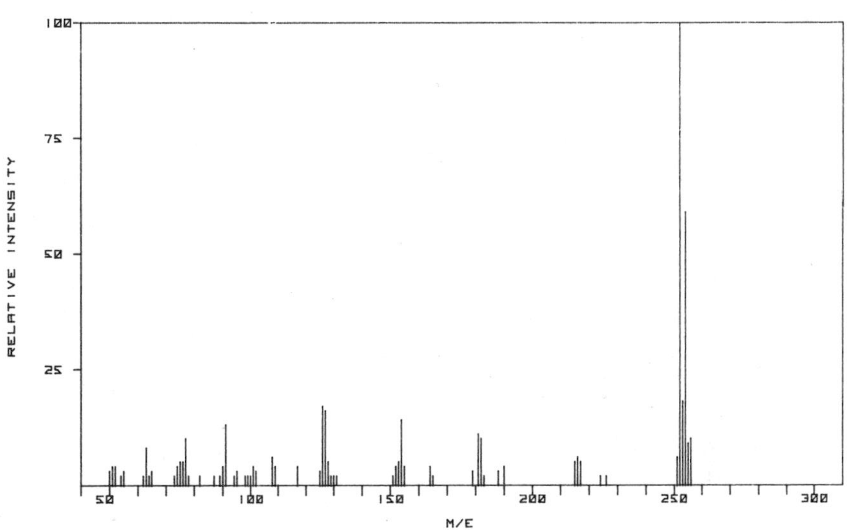

Spectral Data

Mass	Abundance	Mass	Abundance	Mass	Abundance
50	2.6	76	5.0	100	2.3
51	3.5	77	9.6	101	3.8
52	4.1	78	2.0	102	2.9
54	2.0	82	2.3	108	5.6
55	2.6	87	2.3	109	3.8
62	2.3	89	2.3	117	4.4
63	7.6	90	3.5	125	2.6
64	2.3	91	12.6	126	17.3
65	2.6	94	2.3	127	15.5
73	2.3	95	2.6	128	4.7
74	3.5	98	2.3	129	2.0
75	4.7	99	2.3	130	2.0

3,3′-DICHLOROBENZIDINE–*continued*

Spectral Data–*continued*

Mass	Abundance	Mass	Abundance	Mass	Abundance
131	2.0	181	10.8	224	2.3
151	2.3	182	9.9	226	2.0
152	4.4	183	2.3	251	6.1
153	4.7	188	2.9	252	100.0
154	13.7	190	4.1	253	17.8
155	4.1	215	5.0	254	59.1
164	4.4	216	5.8	255	8.8
165	2.3	217	4.7	256	9.6
179	2.6				

Base–neutral extractable
CAS Name: Benzidine, 3,3′-dichloro-
Synonyms

C.I. 23060
Curithane C 126
DCB
4,4′-Diamino-3,3′-dichlorobiphenyl
4,4′-Diamino-3,3′-dichlorodiphenyl
o,o′-Dichlorobenzidine
Dichlorobenzidine base
3,3′-Dichloro-4,4′-biphenyldiamine
3,3′-Dichlorobiphenyl-4,4′-diamine
3,3′-Dichloro-4,4′-diaminobiphenyl
3,3′-Dichloro-4,4′-diaminodiphenyl

CAS Registry No.: 91-94-1
ROTECS Ref.: DD05250
Merck Index Ref.: 3032

Major Ions: 252, 254, 253, 126, 127, 154, 91, 181
EPA Ions: 252, 254, 126

Selected Bibliography

A system for the rapid identification of toxic organic pollutants in water, Donaldson, W. T., Carter, M. H., McGuire, J. M., Comm. Eur. Communities, (Rep.) EUR, 75, (EUR 5360, Proc. Int. Symp. Recent Adv. Assess. Health Eff. Environ. Pollut., Vol. 3,), pp. 1399–1406.

DICHLOROETHYLENES

1,1-Dichloroethylene and *trans*-1,2-dichloroethylene are included in this category.

1,1-Dichloroethylene is an intermediate in the production of the vinylidene polymer plastics Saran and Velon.

trans-1,2-Dichloroethylene is used for retarding fermentation and is a solvent for many substances, including camphor, fats, and phenol.

1,1-DICHLOROETHYLENE $C_2H_2Cl_2$ (96)

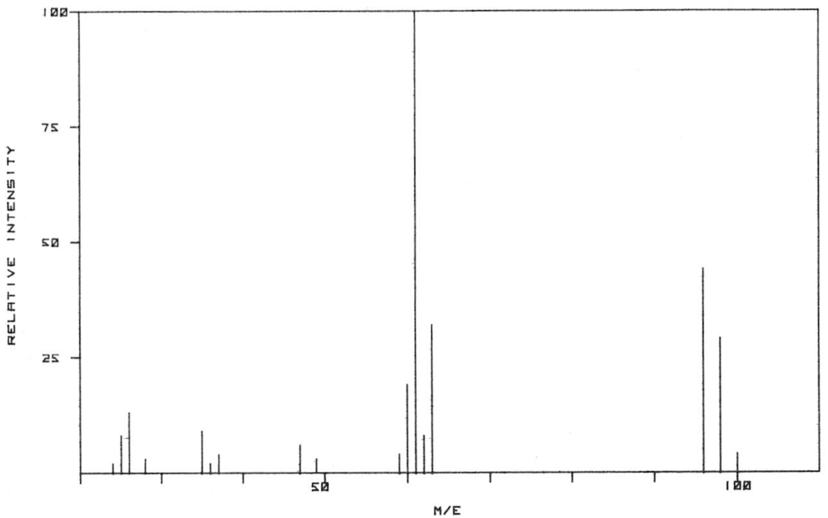

Spectral Data

Mass	Abundance	Mass	Abundance	Mass	Abundance
24	2.1	37	3.5	62	8.4
25	8.4	47	6.0	63	31.9
26	13.4	49	2.8	96	44.3
28	2.5	59	3.9	98	28.9
35	9.2	60	18.9	100	4.1
36	2.3	61	100.0		

1,1-DICHLOROETHYLENE–*continued*

Volatile
CAS Name: Ethylene, 1,1-dichloro-
Synonyms

1,1-DCE	Sconatex
1,1-Dichloroethene	Vinylidene chloride
asym-Dichloroethylene	Vinylidine chloride
NCI-C54262	

CAS Registry No.: 75-35-4
ROTECS Ref.: KV92750
Merck Index Ref.: 9647

Major Ions: 61, 96, 63, 98, 60, 26, 35, 62
EPA Ions: 61, 96, 98

trans-1,2-DICHLOROETHYLENE $C_2H_2Cl_2$ (96)

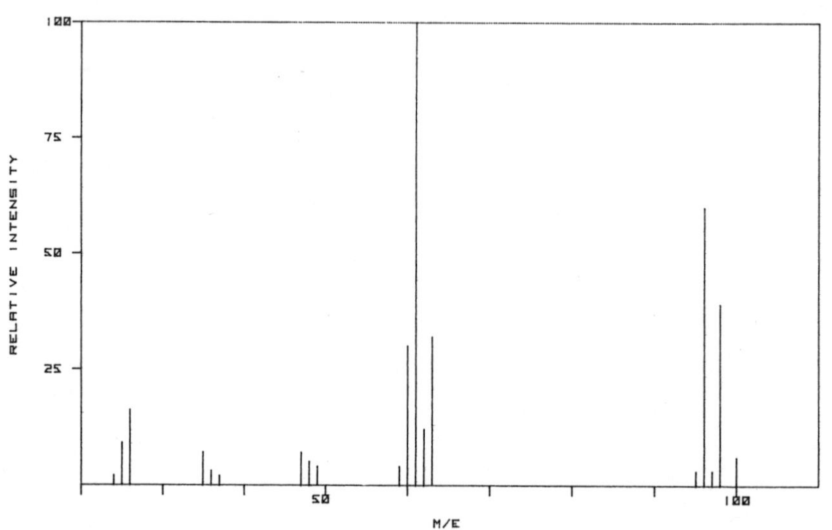

Spectral Data

Mass	Abundance	Mass	Abundance	Mass	Abundance
24	2.0	48	5.1	63	32.3
25	8.6	49	3.7	95	2.9
26	16.3	59	3.8	96	60.2
35	6.5	60	29.7	97	3.1
36	2.8	61	100.0	98	39.3
37	2.2	62	12.2	100	6.1
47	6.2				

Volatile
CAS Name: Ethylene, 1,2-dichloro-, (*E*)-
Synonyms

trans-Acetylene dichlorine
trans-1,2-Dichloroethene
trans-Dichloroethylene
(*E*)-1,2-Dichloroethylene
NCI-C54591

CAS Registry No.: 156-60-5 (formerly 43695-79-0)
ROTECS Ref.: KV94000

Major Ions: 61, 96, 98, 63, 60, 26, 62, 25
EPA Ions: 61, 96, 98

2,4-DICHLOROPHENOL

The benzenesulfonate of this compound is an acaricide, for the control of phytophagous mites.

2,4-DICHLOROPHENOL **$C_6H_4Cl_2O$ (162)**

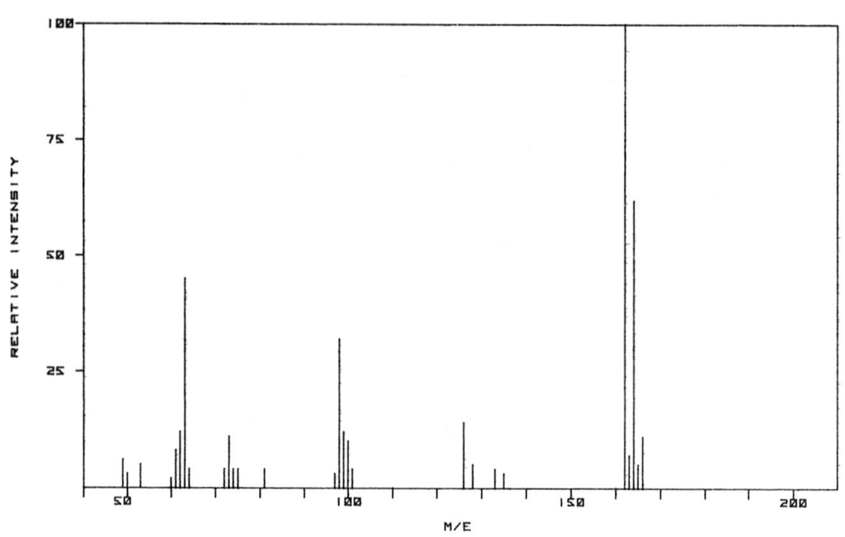

Spectral Data

Mass	Abundance	Mass	Abundance	Mass	Abundance
49	5.7	73	11.0	126	13.9
50	3.2	74	3.9	128	4.6
53	4.6	75	4.2	133	3.5
60	2.1	81	3.5	135	2.5
61	7.6	97	2.9	162	100.0
62	12.2	98	31.5	163	7.3
63	45.4	99	12.3	164	62.2
64	3.7	100	10.3	165	4.5
72	3.7	101	4.0	166	10.5

2,4-DICHLOROPHENOL

Acid extractable
CAS Name: Phenol, 2,4-dichloro-
Synonyms

- DCP
- 4,6-Dichlorophenol
- Isobac

CAS Registry No.: 120-83-2
ROTECS Ref.: SK85750

Major Ions: 162, 164, 63, 98, 126, 99, 62, 73
EPA Ions: 162, 164, 98

DICHLOROPROPANE AND DICHLOROPROPENE

1,2-Dichloropropane and 1,3-dichloropropenes are the compounds included in this category.

1,2-Dichloropropane is an insect fumigant for stored grain. It is a solvent for oils and fats and is used for dry cleaning and degreasing. This compound is also a component (30–35%) of the soil fumigant D-D, used for the control of nematodes.

The fumigant D-D (see above) contains 60–66% of 1,3-dichloropropenes.

1,2-DICHLOROPROPANE $C_3H_6Cl_2$ (112)

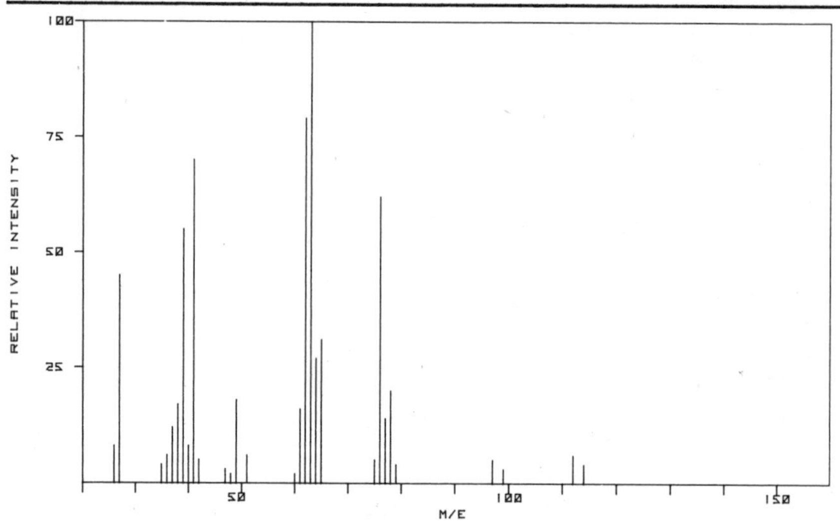

Spectral Data

Mass	Abundance	Mass	Abundance	Mass	Abundance
26	8.1	47	2.9	75	4.7
27	44.7	48	2.2	76	61.7
35	4.2	49	17.9	77	13.7
36	5.9	51	5.8	78	20.2
37	11.7	60	2.3	79	4.0
38	16.6	61	16.5	97	5.2
39	54.9	62	79.2	99	3.0
40	7.7	63	100.0	112	5.8
41	70.1	64	27.0	114	2.8
42	5.4	65	30.8		

Volatile
CAS Name: Propane, 1,2-dichloro-
Synonyms

- α,β-Dichloropropane
- Ent 15,406
- Propylene chloride
- Propylene dichloride
- α,β-Propylene dichloride

CAS Registry No.: 78-87-5 (formerly 26198-64-1)
ROTECS Ref.: TX96250
Merck Index Ref.: 7643

Major Ions: 63, 62, 41, 76, 39, 27, 65, 64
EPA Ions: 63, 65, 112, 114

Selected Bibliography

Negative ion mass spectra of chlorine-containing molecules, Ito, A., Matsumoto, K., Takeuchi, T., *Org. Mass Spectrom.*, **6**(9), 1045–1049 (1972).

Energy partitioning by mass spectrometry. Chloroalkanes and chloroalkenes, Kim, K. C., Beynon, J. H., Cooks, R. G., *J. Chem. Phys.*, **61**(4), 1305–1315 (1974).

A computer algorithm for qualitative identification of mass spectral data acquired in trace level analysis of environmental samples, Davidson, W. C., Smith, M. J., Schaefer, D. J., *Anal. Lett.*, **10**(4), 309–331 (1977).

Direct aqueous injection gas chromatography–mass spectrometry for analysis of organohalides in water at concentrations below the parts per billion level, Fujii, T., *J. Chromatogr.*, **139**(2), 297–302 (1977).

1,3-DICHLOROPROPENE $C_3H_4Cl_2$ (110)

Volatile
CAS Name: Propene, 1,3-dichloro-
Synonyms

- α-Chlorallyl chloride
- γ-Chlorallyl chloride
- 3-Chloropropenyl choride
- 1,3-D
- Dichloropropene
- 1,3-Dichloropropene
- 1,3-Dichloropropene-1
- 1,3-Dichloro-1-propene
- 1,3-Dichloro-2-propene
- α,γ-Dichloropropylene
- 1,3-Dichloropropylene
- Telone
- Telone C
- Telone II

CAS Registry No.: 542-75-6 (formerly 8022-76-2)
ROTECS Ref.: UC831000
Merck Index Ref.: 3051

cis-1,3-DICHLOROPROPENE $C_3H_4Cl_2$ (110)

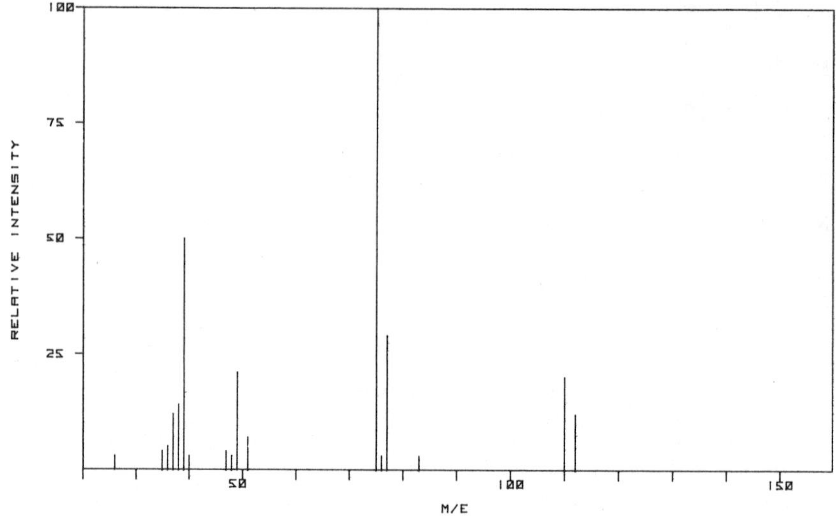

Spectral Data

Mass	Abundance	Mass	Abundance	Mass	Abundance
26	3.1	40	2.7	75	100.0
35	3.7	47	3.6	76	2.9
36	4.6	48	3.2	77	28.9
37	12.1	49	21.2	83	2.6
38	13.9	51	6.5	110	19.6
39	50.3	73	5.9	112	12.2

Volatile

Major Ions: 75, 39, 77, 49, 110, 38, 112, 37
EPA Ions: 75, 77

trans-1,3-DICHLOROPROPENE $C_3H_4Cl_2$ (110)

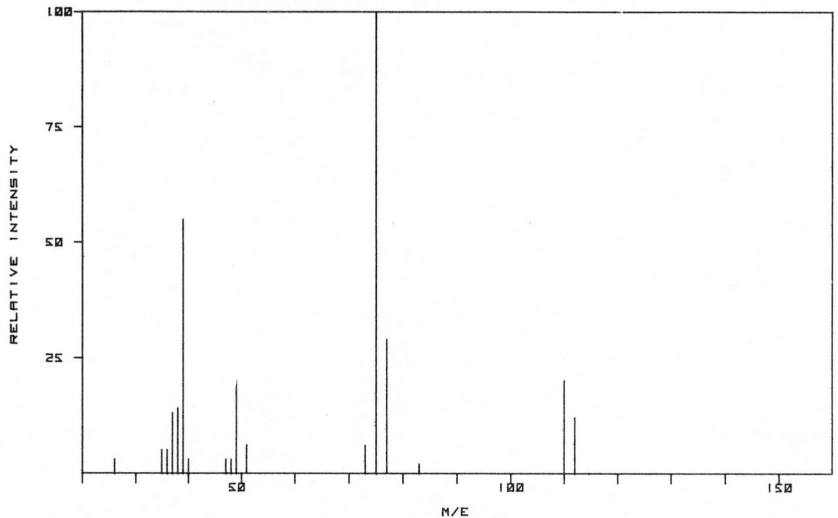

Spectral Data

Mass	Abundance	Mass	Abundance	Mass	Abundance
26	2.9	40	2.7	75	100.0
35	4.6	47	3.2	77	29.1
36	4.5	48	2.8	83	2.2
37	12.8	49	20.4	110	19.8
38	14.4	51	5.8	112	11.7
39	55.0	73	6.2		

Volatile
Major Ions: 75, 39, 77, 49, 110, 38, 37, 112
EPA Ions: 75, 77

2,4-DIMETHYLPHENOL

The dimethylphenols are used in the manufacture of coal tar disinfectants and artificial resins.

2,4-DIMETHYLPHENOL $C_8H_{10}O$ (122)

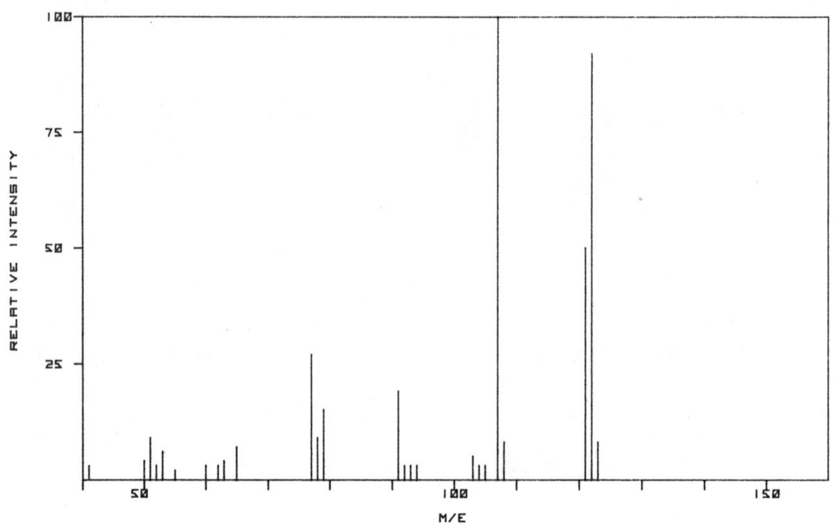

Spectral Data

Mass	Abundance	Mass	Abundance	Mass	Abundance
41	2.9	65	6.8	103	4.8
50	4.0	77	26.9	104	3.4
51	8.9	78	9.0	105	3.1
52	2.9	79	15.0	107	100.0
53	6.3	91	19.4	108	8.0
55	2.2	92	2.5	121	49.6
60	2.7	93	3.0	122	92.1
62	2.6	94	2.5	123	8.1
63	4.2				

2,4-DIMETHYLPHENOL

Acid extractable
CAS Name: 2,4-Xylenol
Synonyms
  4,6-Dimethylphenol
  1-Hydroxy-2,4-dimethylbenzene
  4-Hydroxy-1,3-dimethylbenzene
  m-Xylenol

CAS Registry No.: 105-67-9
ROTECS Ref.: ZE56000
Merck Index Ref.: 9744

Major Ions: 107, 122, 121, 77, 91, 79, 78, 51
EPA Ions: 122, 107, 121

Selected Bibliography

Correlation between electronic fragmentation and the thermal decomposition of toluene and some phenols, Braekman-Danheux, C., Nguyen Cu Quyen, *Ann. Mines Belg.*, (2), 179–184 (1977).

DINITROTOLUENES

2,4- and 2,6-dinitrotoluenes are the compounds included in this category.

2,4-DINITROTOLUENE $C_7H_6N_2O_4$ (182)

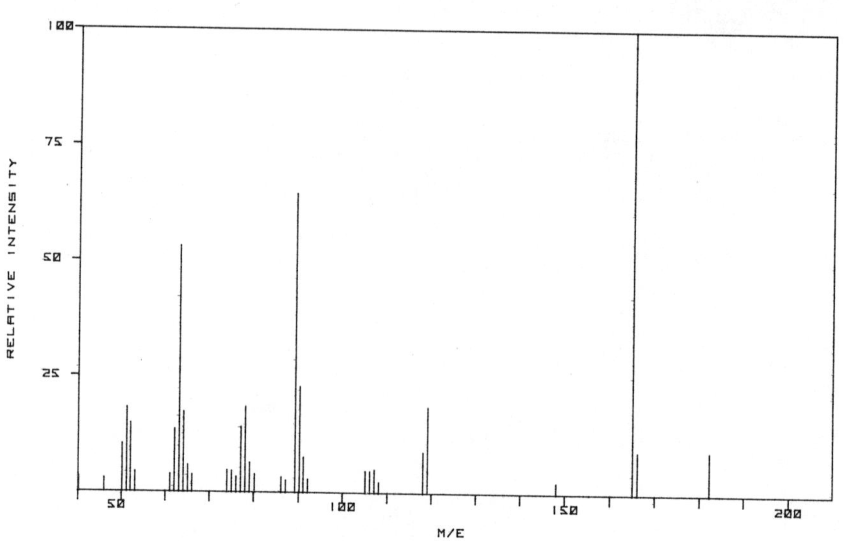

Spectral Data

Mass	Abundance	Mass	Abundance	Mass	Abundance
40	3.2	74	4.6	92	2.7
46	2.8	75	4.4	105	4.6
50	10.2	76	3.3	106	4.5
51	18.1	77	13.9	107	4.9
52	14.7	78	18.4	108	2.3
53	4.3	79	6.3	118	8.7
61	3.8	80	3.8	119	18.5
62	13.4	86	3.1	148	2.4
63	53.0	87	2.5	165	100.0
64	17.2	89	64.4	166	9.1
65	5.7	90	22.7	182	9.1
66	3.7	91	7.5		

Base–neutral extractable
CAS Name: Toluene, 2,4-dinitro-
Synonyms
- 2,4-Dinitrotoluol
- DNT
- 2,4-DNT
- 1-Methyl-2,4-dinitrobenzene
- NCI-C01865

CAS Registry No.: 121-14-2
ROTECS Ref.: XT15750

Major Ions: 165, 89, 63, 90, 119, 78, 51, 64
EPA Ions: 165, 89, 163

Selected Bibliography

Chemical ionization mass spectrometry of explosives, Zitrin, S., Yinon, J., *Adv. Mass Spectrom. Biochem Med.*, **1**, 369–381 (1976).

Feasibility of gunshot residue detection via its organic constituents. Part I. Analysis of smokeless powders by combined gas chromatography–chemical ionization mass spectrometry, Mach, M. H., Pallos, A., Jones, P. F., *J. Forensic Sci.*, **23**(3), 433–445 (1978).

Analysis of explosives using chemical ionization mass spectroscopy, Pate, C. T., Mach, M. H., *Int. J. Mass Spectrom. Ion Phys.*, **26**(3), 267–277 (1978).

Ionization of nitrotoluene compounds in negative ion plasma chromatography, Spangler, G. E., Lawless, P. A., *Anal. Chem.*, **50**(7), 884–892 (1978).

2,6-DINITROTOLUENE $C_7H_6N_2O_4$ (182)

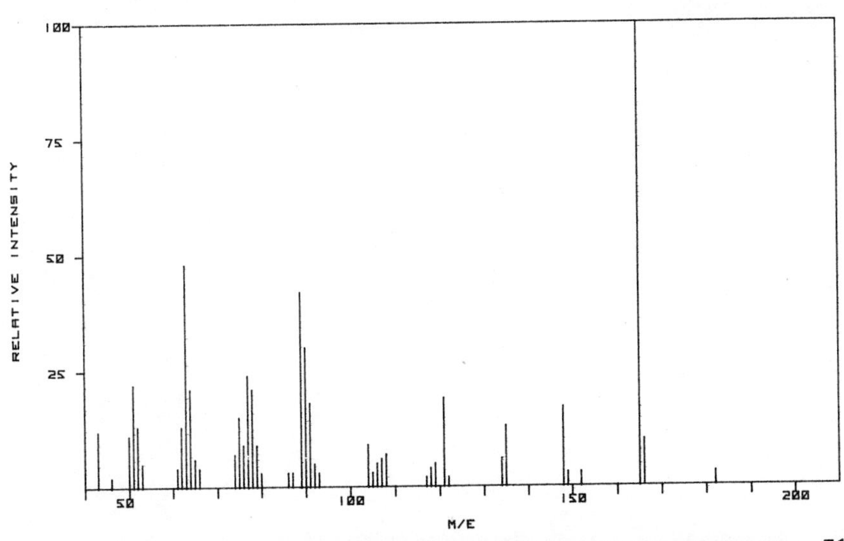

2,6-DINITROTOLUENE–*continued*

Spectral Data

Mass	Abundance	Mass	Abundance	Mass	Abundance
43	11.6	77	23.9	108	6.9
46	2.2	78	20.8	117	2.2
50	11.3	79	9.4	118	4.3
51	22.2	80	2.6	119	4.5
52	13.1	86	2.6	121	18.7
53	5.1	87	3.4	122	2.0
61	4.3	89	41.7	134	6.3
62	13.1	90	30.4	135	13.4
63	47.7	91	18.2	148	17.0
64	21.0	92	4.9	149	2.6
65	5.7	93	2.5	152	3.4
66	3.9	104	8.5	165	100.0
74	6.6	105	3.2	166	9.7
75	14.8	106	4.8	182	2.5
76	9.3	107	5.6		

Base–neutral extractable
CAS Name: Toluene, 2,6-dinitro-
Synonym
    2-Methyl-1,3-dinitrobenzene

CAS Registry No.: 606-20-2
ROTECS Ref.: XT19250

Major Ions: 165, 63, 89, 90, 77, 51, 64, 78
EPA Ions: 165, 63, 121

1,2-DIPHENYLHYDRAZINE

1,2-Diphenylhydrazine decomposes to afford azobenzene during gas chromatography. The mass spectrum of the latter compound is reported here.

1,2-DIPHENYLHYDRAZINE $C_{12}H_{12}N_2$ (184)

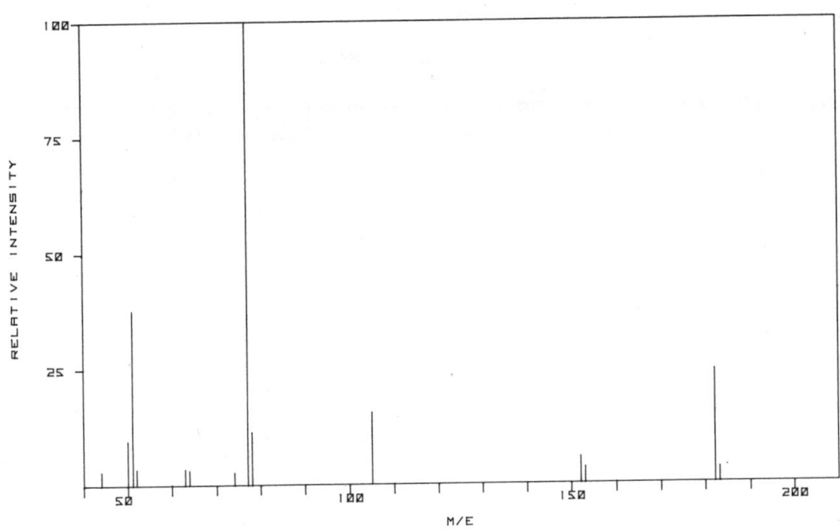

Spectral Data for Azobenzene (see above)

Mass	Abundance	Mass	Abundance	Mass	Abundance
44	2.8	64	3.1	152	5.7
50	9.5	74	2.6	153	3.4
51	37.6	77	100.0	182	24.2
52	3.4	78	11.3	183	3.1
63	3.4	105	15.5		

1,2-DIPHENYLHYDRAZINE–*continued*

Base–neutral extractable
CAS Name: Hydrazobenzene
Synonyms
N,N'-Bianiline
N,N'-Diphenylhydrazine
1,1'-Hydrazobisbenzene
Hydrazodibenzene

CAS Registry No.: 122-66-7
ROTECS Ref.: MW26250

Major Ions: 77, 51, 182, 105, 78, 50, 152, 153 }
EPA Ions: 77, 93, 105 } for azobenzene (see p. 73).

Selected Bibliography

Mass spectral analysis of medicinal pyrazolidinediones, Locock, R. A., Moskalyk, R. E., Chatten, L. G., Lundy, L. M., *J. Pharm. Sci.*, **63**(12), 1896–1901 (1974).

ENDOSULFAN AND METABOLITES

α-Endosulfan, β-endosulfan, and endosulfan sulfate are the compounds included in this category.

Endosulfan (a broad-spectrum contact insecticide) is a mixture of the α- and β-isomers. It is highly toxic to fish, and was responsible for a massive fish kill in the Rhine River. This compound is rapidly metabolized to the sulfate.

ENDOSULFAN — $C_9H_6Cl_6O_3S$ (404)

Pesticide

CAS Name: 5-Norbornene-2,3-dimethanol, 1,4,5,6,7,7-hexachloro-, cyclic sulfite

Synonyms

- Benzoepin
- Beosit
- Bio 5,462
- Chlorthiepin
- Cyclodan
- Endosulfan 35EC
- Endosulphan
- Ent 23,979
- FMC 5462
- α,β-1,2,3,4,7,7-Hexachlorobicyclo(2,2,1)-2-heptene-5,6-bisoxymethylene sulfite
- 1,2,3,4,7,7-Hexachlorobicyclo(2,2,1)-hepten-5,6-bioxymethylene sulfite
- Hexachlorohexahydromethano-2,4,3-benzodioxathiepin-3-oxide
- 6,7,8,9,10,10-Hexachloro-1,5,5α,6,9,9α-hexahydro-6,9-methano-2,4,3-benzodioxathiepin-3-oxide
- 1,4,5,6,7,7-Hexachloro-5-norbornene-2,3-dimethanol cyclic sulfite
- HOE 2,671
- Insectophene
- Kop-thiodan
- Malix
- NCI-C00566
- NIA-5462
- Niagara 5,462
- OMS 570
- SD 4314
- Sulfurous acid, cyclic ester with 1,4,5,6,7,7-hexachloro-5-norbornene-2,3-dimethanol
- Thifor
- Thimul
- Thiodan
- Thiofor
- Thionex
- Thiosulfan
- Tionel
- Tiovel

CAS Registry No.: 115-29-7 (formerly 8003-45-0)

ROTECS Ref.: RB92750

Merck Index Ref.: 3519

α-ENDOSULFAN $C_9H_6Cl_6O_3S$ (404)

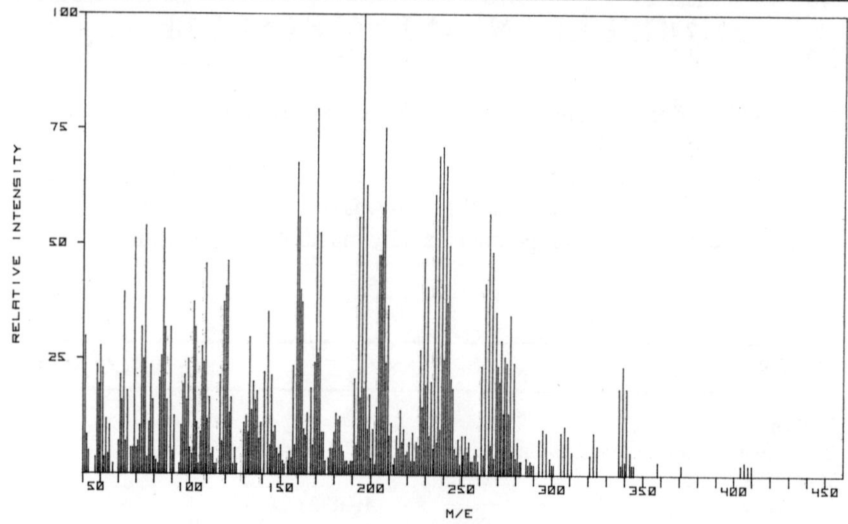

Spectral Data

Mass	Abundance	Mass	Abundance	Mass	Abundance
40	7.6	72	10.3	100	5.5
41	29.7	73	31.7	101	4.1
42	8.3	74	24.8	102	37.2
43	4.8	75	53.8	103	31.7
47	3.4	76	3.4	104	11.0
48	23.4	77	11.0	105	2.1
49	19.3	78	23.4	106	9.0
50	27.6	79	15.9	107	27.6
51	22.8	80	3.4	108	24.1
52	3.4	81	2.8	109	45.5
53	11.7	82	2.1	110	11.7
54	4.1	83	20.7	111	16.6
55	10.3	84	25.5	112	4.1
57	2.1	85	53.1	113	5.5
60	6.9	86	31.7	114	2.1
61	21.4	87	15.9	115	2.1
62	15.9	89	31.7	117	21.4
63	39.3	90	4.8	118	6.9
64	6.9	91	12.4	119	37.2
65	17.9	94	2.1	120	40.7
67	5.5	95	10.3	121	46.2
68	5.5	96	19.3	122	13.1
69	51.0	97	21.4	123	16.6
70	5.5	98	15.9	124	2.1
71	6.9	99	24.8	125	5.5

Spectral Data–*continued*

Mass	Abundance	Mass	Abundance	Mass	Abundance
126	2.1	182	11.7	230	19.3
130	11.0	183	12.4	231	40.7
131	12.4	184	6.2	232	8.3
132	9.0	185	4.8	233	20.0
133	29.7	186	2.8	234	5.5
134	13.8	187	2.8	235	60.7
135	20.0	188	2.1	236	6.9
136	15.9	189	2.8	237	69.0
137	17.9	190	2.8	238	9.7
138	7.6	191	20.7	239	71.0
139	11.0	192	6.2	240	24.8
141	22.1	193	55.9	241	66.9
143	35.2	194	16.6	242	37.2
144	6.2	195	100.0	243	49.7
145	21.4	196	18.6	244	20.7
146	9.0	197	62.8	245	18.6
147	10.3	198	9.7	246	5.5
148	5.5	199	17.2	247	4.1
149	4.1	200	3.4	248	7.6
150	6.2	201	9.7	249	2.1
151	2.8	202	2.1	250	8.3
152	2.1	203	14.5	251	4.1
154	2.8	204	47.6	252	8.3
155	4.8	205	47.6	253	4.8
156	3.4	206	57.9	254	6.9
157	23.4	207	75.2	255	2.8
158	6.2	208	24.1	256	2.8
159	67.6	209	36.6	257	4.1
160	55.9	210	8.3	258	5.5
161	40.0	211	11.0	259	2.8
162	37.2	212	2.1	261	23.4
163	9.7	213	3.4	262	4.1
164	8.3	214	8.3	263	41.4
165	13.1	215	5.5	265	56.6
167	18.6	216	13.8	266	6.2
168	6.2	217	6.9	267	48.3
169	24.1	218	9.7	268	3.4
170	79.3	219	4.1	269	35.2
171	26.2	220	4.8	270	23.4
172	52.4	221	6.9	271	20.0
173	9.0	222	2.8	272	29.0
174	9.0	223	9.0	273	13.1
175	2.8	224	2.8	274	25.5
177	3.4	225	6.9	275	24.1
178	5.5	226	6.2	276	13.1
179	5.5	227	26.9	277	34.5
180	9.0	228	14.5	278	4.8
181	13.1	229	46.9	279	24.1

α-ENDOSULFAN–*continued*

Spectral Data–*continued*

Mass	Abundance	Mass	Abundance	Mass	Abundance
280	3.4	299	3.4	339	23.4
281	6.9	300	2.1	340	2.8
282	2.8	301	2.1	341	18.6
283	2.8	305	9.0	343	4.8
286	3.4	307	10.3	344	2.1
287	2.1	309	8.3	345	2.1
288	2.8	311	4.8	358	2.8
289	2.1	321	4.1	371	2.1
290	2.1	323	9.0	404	2.1
293	7.6	325	6.2	406	2.8
295	9.7	337	18.6	408	2.1
297	9.0	338	2.1	410	2.1

Pesticide

CAS Name: 5-Norbornene-2,3-dimethanol, 1,4,5,6,7,7-hexachloro-, cyclic sulfite, *endo*

Synonyms

Endosulfan I

(3α,5aβ,6α,9α,9aβ)-6,7,8,9,10,10-Hexachloro-1,5,5a,6,9,9a-hexahydro-6,9-methano-2,4,3-benzodioxathiepin 3-oxide

α-Thiodan

CAS Registry No.: 959-98-8 (formerly 12640-58-3, 19595-59-6, 29106-31-8, 33213-66-0)

Major Ions: 195, 170, 207, 239, 237, 159, 241, 197
EPA Ions: 201, 283, 278

β-ENDOSULFAN $C_9H_6Cl_6O_3S$ (404)

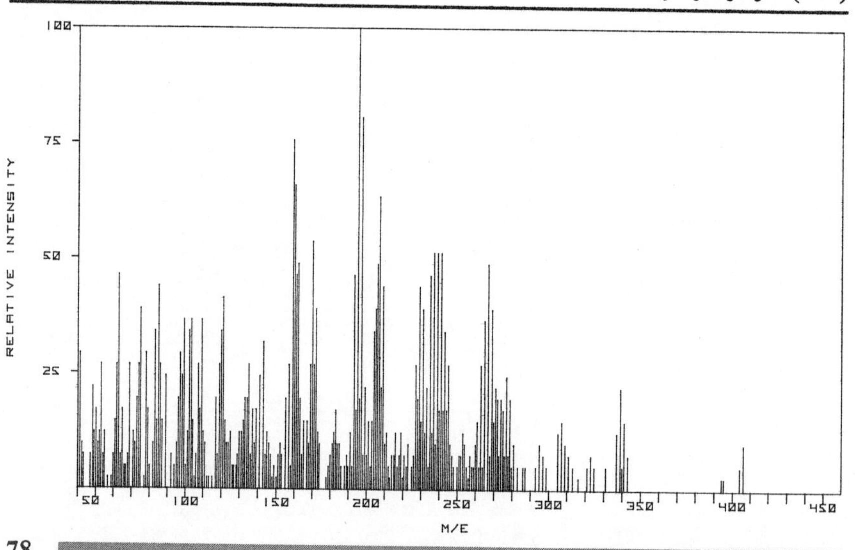

Spectral Data

Mass	Abundance	Mass	Abundance	Mass	Abundance
40	12.2	100	4.9	153	7.3
41	29.3	101	12.2	155	19.5
42	9.8	102	34.1	157	26.8
43	7.3	103	36.6	158	4.9
46	7.3	104	14.6	159	75.6
48	22.0	105	2.4	160	65.9
49	12.2	106	7.3	161	46.3
50	17.1	107	26.8	162	48.8
51	9.8	108	17.1	163	19.5
52	12.2	109	36.6	164	7.3
53	26.8	110	12.2	165	14.6
54	7.3	111	9.8	167	14.6
55	12.2	112	2.4	168	9.8
57	2.4	113	2.4	169	26.8
59	2.4	115	2.4	170	53.7
60	7.3	117	19.5	171	26.8
61	14.6	118	7.3	172	39.0
62	26.8	119	26.8	173	12.2
63	46.3	120	34.1	174	9.8
64	7.3	121	41.5	178	2.4
65	17.1	122	14.6	179	4.9
66	4.9	123	9.8	180	7.3
67	4.9	124	9.8	181	9.8
68	7.3	125	12.2	182	12.2
69	26.8	126	4.9	183	17.1
71	12.2	127	4.9	184	9.8
72	9.8	129	7.3	185	9.8
73	19.5	130	12.2	186	4.9
74	26.8	131	12.2	188	4.9
75	39.0	132	14.6	189	7.3
77	2.4	133	19.5	190	4.9
78	29.3	134	19.5	191	12.2
79	17.1	135	26.8	192	4.9
80	4.9	136	7.3	193	46.3
82	9.8	137	17.1	194	17.1
83	34.1	138	9.8	195	100.0
84	14.6	139	17.1	196	19.5
85	43.9	141	24.4	197	80.5
86	26.8	143	31.7	198	7.3
87	14.6	144	7.3	199	22.0
89	24.4	145	12.2	200	7.3
92	7.3	146	9.8	201	14.6
94	4.9	147	7.3	202	4.9
95	9.8	148	2.4	203	14.6
96	19.5	149	4.9	204	34.1
97	29.3	150	2.4	205	39.0
98	24.4	151	7.3	206	48.8
99	36.6	152	9.8	207	63.4

β-ENDOSULFAN–*continued*

Spectral Data–*continued*

Mass	Abundance	Mass	Abundance	Mass	Abundance
208	22.0	242	17.1	276	7.3
209	43.9	243	34.1	277	24.4
210	9.8	244	17.1	278	7.3
211	12.2	245	26.8	279	19.5
212	4.9	246	9.8	280	4.9
213	2.4	247	7.3	281	9.8
214	7.3	248	4.9	283	4.9
215	7.3	250	4.9	286	4.9
216	12.2	251	7.3	287	4.9
217	4.9	252	7.3	293	4.9
218	7.3	253	12.2	295	7.3
219	12.2	254	9.8	297	7.3
220	2.4	255	4.9	299	4.9
221	7.3	256	2.4	305	12.2
222	4.9	257	7.3	307	14.6
223	9.8	258	4.9	309	9.8
225	4.9	259	4.9	311	7.3
226	7.3	260	9.8	313	4.9
227	26.8	261	14.6	316	2.4
228	19.5	262	4.9	321	4.9
229	43.9	263	26.8	323	7.3
230	14.6	264	4.9	325	4.9
231	39.0	265	36.6	331	4.9
232	12.2	267	48.8	337	12.2
233	22.0	268	7.3	339	22.0
234	4.9	269	39.0	340	4.9
235	46.3	270	14.6	341	14.6
236	12.2	271	22.0	343	7.3
237	51.2	272	19.5	394	12.4
238	9.8	273	7.3	395	2.4
239	51.2	274	19.5	404	4.9
240	17.1	275	17.1	406	9.8
241	51.2				

Pesticide

CAS Name: 5-Norbornene-2,3-dimethanol, 1,4,5,6,7,7-hexachloro-, cyclic sulfite, *exo*-

Synonyms

- Endosulfan II
- General Weed Killer
- (3α,5aα,6β,9β,9aα)-6,7,8,9,10,10-Hexachloro-1,5,5a,6,9,9a-hexahydro-6,9-methano-2,4,3-benzodioxathiepin 3-oxide
- β-Thiodan

CAS Registry No.: 33213-65-9 (formerly 891-86-1, 12640-59-4, 19670-15-6)

Major Ions: 195, 197, 159, 160, 207, 170, 239, 237

EPA Ions: 201, 283, 278

ENDOSULFAN SULFATE $C_9H_6Cl_6O_4S$ (420)

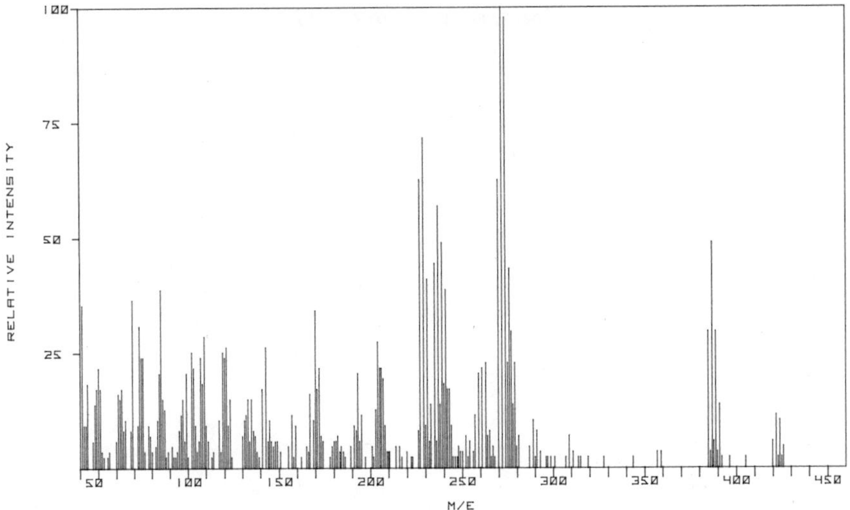

Spectral Data

Mass	Abundance	Mass	Abundance	Mass	Abundance
41	35.2	75	23.9	103	21.6
42	9.1	76	3.4	104	9.1
43	9.1	78	9.1	105	3.4
44	18.2	79	6.8	106	5.7
47	5.7	80	3.4	107	23.9
48	13.6	82	4.5	108	18.2
49	17.0	83	10.2	109	28.4
50	21.6	84	20.5	110	9.1
51	17.0	85	38.6	111	5.7
52	3.4	86	14.8	113	2.3
53	2.3	87	12.5	114	3.4
55	2.3	88	2.3	117	10.2
56	3.4	89	3.4	118	3.4
60	5.7	91	4.5	119	25.0
61	15.9	92	2.3	120	23.9
62	14.8	93	2.3	121	26.1
63	17.0	94	3.4	122	9.1
64	8.0	95	8.0	123	14.8
65	10.2	96	11.4	124	2.3
68	8.0	97	14.8	130	6.8
69	36.4	98	5.7	131	10.2
72	9.1	99	20.5	132	11.4
73	30.7	100	2.3	133	14.8
74	23.9	102	25.0	134	5.7

ENDOSULFAN SULFATE–*continued*

Spectral Data–*continued*

Mass	Abundance	Mass	Abundance	Mass	Abundance
135	14.8	201	4.5	259	20.5
136	8.0	202	2.3	261	21.6
137	6.8	203	12.5	263	22.7
138	3.4	204	27.3	264	6.8
139	2.3	205	21.6	265	8.0
141	17.0	206	21.6	266	2.3
143	26.1	207	19.3	267	4.5
144	5.7	208	9.1	268	2.3
145	10.2	209	3.4	270	62.5
146	5.7	210	3.4	272	100.0
147	4.5	214	4.5	274	97.7
148	5.7	216	4.5	275	22.7
149	5.7	217	2.3	276	43.2
151	3.4	220	3.4	277	29.5
155	4.5	222	2.3	278	13.6
157	11.4	223	2.3	279	22.7
158	2.3	226	8.0	280	4.5
159	9.1	227	62.5	281	6.8
162	2.3	229	71.6	287	4.5
165	4.5	230	9.1	289	10.2
166	3.4	231	40.9	290	2.3
167	15.9	232	5.7	291	8.0
169	10.2	233	13.6	293	3.4
170	34.1	235	44.3	296	2.3
171	17.0	236	5.7	297	2.3
172	21.6	237	56.8	299	2.3
173	6.8	238	13.6	301	2.3
174	5.7	239	48.9	307	2.3
178	2.3	240	18.2	309	6.8
179	4.5	241	38.6	311	3.4
180	5.7	242	17.0	314	2.3
181	5.7	243	17.0	315	2.3
182	6.8	244	9.1	319	2.3
183	3.4	245	2.3	328	2.3
184	4.5	246	2.3	344	2.3
185	3.4	247	2.3	357	3.4
186	2.3	248	4.5	359	3.4
189	4.5	249	3.4	385	29.5
191	9.1	250	3.4	386	3.4
192	8.0	252	6.8	387	48.9
193	20.5	253	2.3	388	5.7
194	5.7	254	5.7	389	29.5
195	11.4	256	3.4	390	3.4
197	2.3	257	11.4	391	13.6

Spectral Data–*continued*

Mass	Abundance	Mass	Abundance	Mass	Abundance
392	2.3	420	5.7	424	10.2
396	2.3	422	11.4	425	2.3
405	5.7	423	2.3	426	4.5

Pesticide
CAS Name: 5-Norbornene-2,3-dimethanol, 1,4,5,6,7,7-hexachloro-, cyclic sulfate
Synonyms

6,7,8,9,10,10-Hexachloro-1,5,5a,6,9,9a-hexahydro-6,9-methano-2,4,3-benzodioxathiepin 3,3-dioxide
Thiodan sulfate

CAS Registry No.: 1031-07-8 (formerly 6749-25-3)

Major Ions: 272, 274, 229, 270, 227, 237, 387, 239
EPA Ions: 272, 387, 422

ENDRIN AND METABOLITES

Endrin and endrin aldehyde are the compounds included in this category.

Endrin, a powerful insecticide, is a stereoisomer of dieldrin (*q.v.*). It is oxidized to endrin aldehyde.

ENDRIN $C_{12}H_8Cl_6O$ (378)

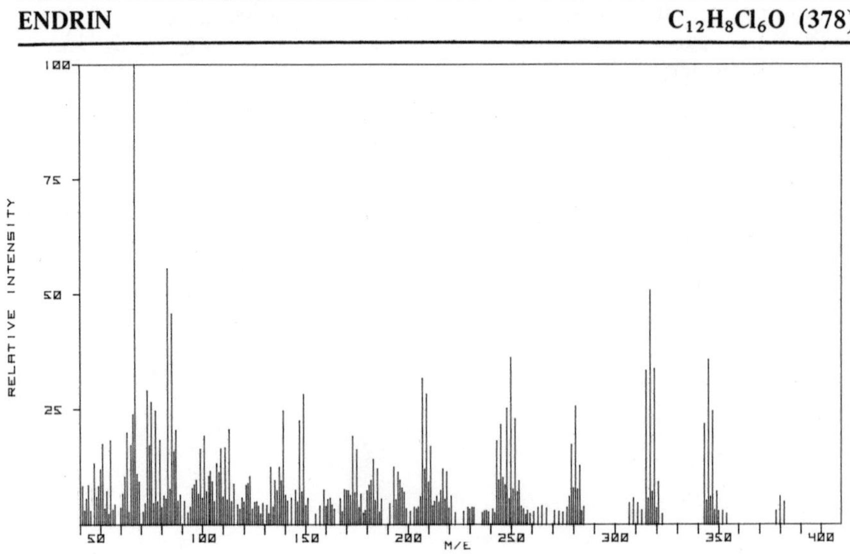

Spectral Data

Mass	Abundance	Mass	Abundance	Mass	Abundance
41	8.2	60	3.5	77	24.6
42	2.9	61	6.5	78	4.8
43	5.5	62	10.3	79	18.3
44	8.4	63	19.9	80	3.6
45	2.8	64	2.5	81	6.1
47	13.2	65	17.1	82	5.4
48	5.9	66	23.8	83	55.6
49	8.1	67	100.0	84	7.6
50	11.9	68	10.9	85	45.7
51	17.4	69	9.2	86	15.7
52	3.3	71	2.6	87	20.4
53	7.1	72	4.4	88	4.9
54	2.2	73	28.9	89	6.3
55	18.2	74	17.1	91	4.9
56	3.0	75	26.5	93	2.4
57	4.2	76	4.4	94	3.6

Spectral Data–*continued*

Mass	Abundance	Mass	Abundance	Mass	Abundance
95	7.7	147	22.5	208	11.9
96	8.5	148	6.9	209	28.2
97	9.6	149	28.2	210	9.1
98	6.5	150	4.0	211	16.9
99	16.3	151	5.6	212	3.9
100	5.6	155	2.2	213	4.9
101	19.1	157	3.9	214	6.0
102	7.0	159	7.4	215	4.7
103	10.4	160	3.8	216	7.2
104	11.5	161	5.3	217	12.0
105	9.2	162	5.6	218	5.3
106	4.8	163	4.1	219	11.3
107	13.2	164	3.2	220	3.4
108	11.2	167	5.5	221	6.0
109	16.4	168	2.6	223	2.4
110	5.9	169	7.5	227	2.7
111	16.6	170	7.3	229	3.6
112	5.2	171	7.2	230	3.2
113	20.6	172	6.2	231	3.6
114	4.9	173	19.1	232	3.5
115	8.7	174	6.7	236	2.4
117	4.2	175	16.2	237	2.8
118	3.2	176	3.5	238	2.9
119	5.7	177	6.5	239	2.6
120	4.8	178	2.3	241	3.2
121	8.4	179	2.9	242	2.4
122	8.9	180	7.2	243	18.0
123	10.4	181	8.4	244	9.5
124	3.2	182	9.5	245	21.5
125	4.7	183	14.0	246	10.1
126	4.8	184	5.1	247	8.5
127	3.9	185	12.0	248	25.1
128	2.1	186	2.5	249	5.4
129	4.5	187	5.4	250	36.1
131	4.1	191	4.4	251	7.6
132	2.2	193	12.4	252	22.8
133	12.4	194	5.2	253	7.0
134	3.7	195	11.3	254	9.4
135	9.5	196	9.6	255	3.8
136	7.2	197	7.8	256	3.2
137	12.3	198	6.8	257	2.0
138	9.4	199	3.3	258	2.9
139	24.6	201	2.7	259	2.2
140	6.2	203	3.6	261	2.6
141	5.0	204	3.2	263	3.5
143	5.6	205	3.6	265	3.9
145	7.3	206	5.9	267	3.4
146	4.8	207	31.6	271	2.8

ENDRIN–*continued*

Spectral Data–*continued*

Mass	Abundance	Mass	Abundance	Mass	Abundance
273	2.7	309	5.6	344	5.1
275	2.5	311	4.5	345	35.6
277	3.5	313	3.0	346	5.9
278	5.9	315	33.3	347	24.5
279	17.3	316	5.5	348	3.0
280	7.8	317	50.8	349	7.1
281	25.5	318	7.0	350	2.8
282	7.6	319	33.7	352	2.8
283	12.7	320	3.4	354	2.2
284	2.8	321	9.1	378	2.8
285	3.7	323	2.1	380	6.0
307	4.5	343	21.7	382	4.8

Pesticide

CAS Name: 1,4:5,8-Dimethanonaphthalene, 1,2,3,4,10,10-hexachloro-6,7-epoxy-1,4,4a,5,6,7,8,8a-octahydro-, *endo,endo-*

Synonyms

- Compound 269
- EN 57
- Endrex
- Endricol
- Ent 17,251
- Experimental insecticide 269
- Hexachloroepoxyoctahydro-*endo,endo*-dimethanonaphthalene
- Hexadrin
- Mendrin
- NCI-C00157
- Nendrin
- Oktanex

CAS Registry No.: 72-20-8 (formerly, 8072-14-8, 12715-94-5, 16386-25-7, 16850-17-2, 17578-52-8, 18466-93-8, 25320-73-4)

ROTECS Ref.: IO15750

Merck Index Ref.: 3522

Major Ions: 67, 83, 317, 85, 250, 345, 319, 315

EPA Ions: 81, 263, 82

Selected Bibliography

Applications of mass spectrometry to trace determinations of environmental toxic materials, Abramson, F. P., *Anal. Chem.*, **44**(14), 28A–33A,35A (1972).

Positive chemical ionization mass spectra of polycyclic chlorinated pesticides, Biros, F. J., Dougherty, R. C., Dalton, J., *Org. Mass Spectrom.*, **6**(11), 1161–1169 (1972).

Negative chemical ionization mass spectra of polycyclic chlorinated insecticides, Dougherty, R. C., Dalton, J., Biros, F. J., *Org. Mass Spectrom.*, **6**(11), 1171–1181 (1972).

Isolation and characterization of some methanonaphthalene photoproducts, Onuska, F. I., Comba, M. E., *Biomed. Mass Spectrom.*, **2**(4), 169–175 (1975).

Application of coupled gas chromatography–mass spectrometry in methods for the study and determination of pesticide residues and organic micropollutants in environmental and food materials, Mestres, R., Chevallier, C., Espinoza, C., Cornet, R., *Ann. Falsif. Exper. Chim.*, **70**(751), 177–188 (1977).

ENDRIN ALDEHYDE $C_{12}H_8Cl_6O$ (378)

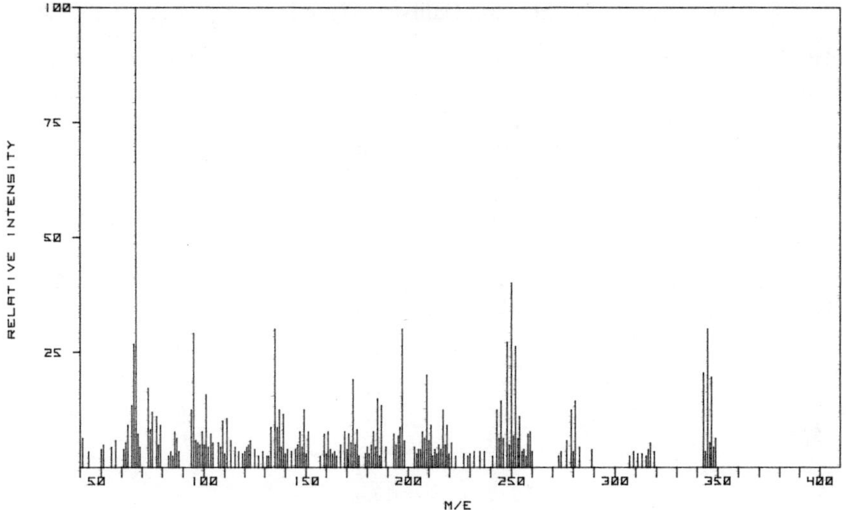

Spectral Data

Mass	Abundance	Mass	Abundance	Mass	Abundance
41	6.2	86	7.6	120	3.3
44	3.3	87	6.2	121	4.3
50	3.8	88	3.3	122	4.8
51	4.8	94	12.4	123	5.7
55	4.3	95	29.0	125	3.8
57	5.7	96	5.7	127	2.4
61	3.8	97	5.2	129	3.3
62	5.2	98	4.8	131	2.4
63	9.0	99	7.6	132	2.4
65	13.3	100	4.8	133	8.6
66	26.7	101	15.7	135	30.0
67	100.0	102	4.3	136	8.6
68	7.1	103	7.1	137	12.4
69	4.3	104	5.2	138	4.3
73	17.1	107	5.2	139	11.4
74	8.1	108	4.3	140	2.9
75	11.9	109	10.0	141	3.8
77	11.0	110	2.9	143	3.3
78	4.8	111	10.5	145	3.8
79	9.0	113	5.7	146	4.8
83	2.4	115	4.3	147	7.6
84	3.3	117	3.3	148	4.3
85	2.4	119	2.9	149	12.4

ENDRIN ALDEHYDE–*continued*

Spectral Data–*continued*

Mass	Abundance	Mass	Abundance	Mass	Abundance
150	2.9	198	5.7	250	40.0
151	7.6	203	4.3	251	6.7
157	2.4	204	2.9	252	26.2
159	7.1	205	3.8	253	6.2
160	2.9	206	3.8	254	11.0
161	7.6	207	7.6	255	3.3
162	3.8	208	6.2	256	3.8
163	2.9	209	20.0	257	2.4
164	3.3	210	5.7	258	7.1
165	2.4	211	9.0	259	7.6
167	4.8	212	2.4	260	3.3
169	7.6	213	3.8	273	2.4
170	3.8	214	2.9	274	3.3
171	7.1	215	4.8	277	5.7
172	5.2	216	3.8	279	12.4
173	19.0	217	12.4	280	3.3
174	4.8	218	4.8	281	14.3
175	8.1	219	9.0	283	4.3
176	2.4	220	2.9	289	3.8
179	2.9	221	5.2	307	2.4
180	4.3	223	2.4	309	3.3
181	2.9	227	2.9	311	2.9
182	4.8	229	2.4	313	2.9
183	7.6	230	2.9	315	2.4
184	4.3	232	3.3	316	3.8
185	14.8	235	3.3	317	5.2
186	2.4	237	3.3	319	3.3
187	13.3	241	2.4	343	20.5
189	4.3	243	12.4	344	3.3
193	7.1	244	6.2	345	30.0
194	4.8	245	14.3	346	5.2
195	6.7	246	6.2	347	19.5
196	8.6	248	27.1	348	4.3
197	30.0	249	4.8	349	6.2

Pesticide

CAS Name: 1,2,4-Methenocyclopenta[*cd*]pentalene-5-carboxaldehyde, 2,2a,3,3,4,7-hexachlorodecahydro, (1α,2β,2aβ,4β, 4aβ,5α,6aβ,6bβ,7*R**)-

Synonym

SD 7442

CAS Registry No.: 7421-93-4

Major Ions: 67, 250, 345, 197, 135, 95, 248, 26

EPA Ions: (not listed)

Selected Bibliography

Endrin transformations in soil, Nash, R. G., Beall, M. L., Jr., Harris, W. G., *J. Environ. Qual.*, **1**(4), 391–394 (1972).

Pesticide residue levels in soils, fiscal year 1969. National Soils Monitoring Program, Wiersma, G. B., Tai, H., Sand, P. F., *Pestic. Monit. J.*, **6**(3), 194–228 (1972).

Gas chromatographic identifications of some organochlorine pesticides and their photoalteration products by means of Kovats' retention indices, Onuska, F. I., Comba, M. E., *J. Chromatogr.*, **119**, 385–399 (1976).

Structure of endrin aldehyde, Bird, C. W., Khan, R., Richardson, A. C., *Chem. Ind. (London)*, (7), 231–232 (1978).

ETHYLBENZENE

This compound is a major petroleum constituent.

ETHYLBENZENE C_8H_{10} **(106)**

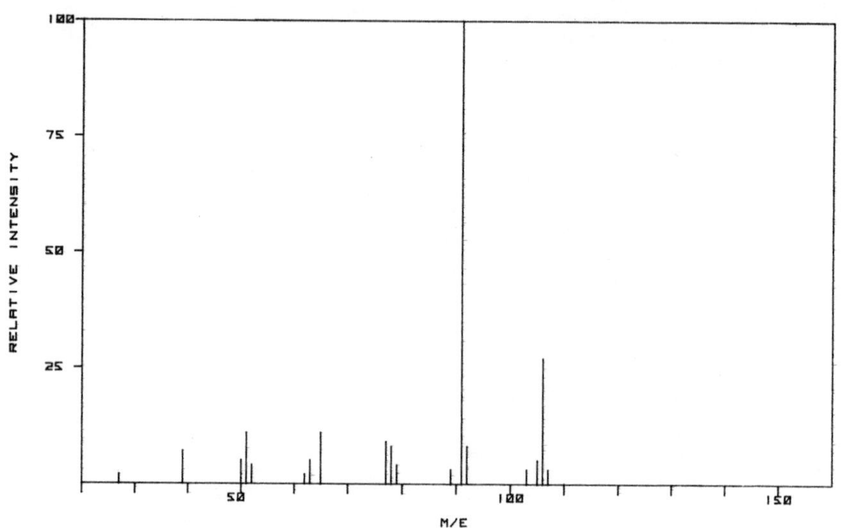

Spectral Data

Mass	Abundance	Mass	Abundance	Mass	Abundance
27	2.2	63	4.9	91	100.0
39	6.5	65	10.9	92	7.6
50	5.3	77	9.2	103	3.1
51	10.7	78	8.0	105	4.5
52	3.5	79	4.2	106	27.3
62	2.0	89	2.8	107	2.5

Volatile
CAS Name: Benzene, ethyl-
Synonyms
- EB
- Ethyl benzene
- Ethylbenzol
- Phenylethane

ETHYLBENZENE

CAS Registry No.: 100-41-4
ROTECS Ref.: DA07000
Merck Index Ref.: 3695

Major Ions: 91, 106, 65, 51, 77, 78, 92, 39
EPA Ions: 91, 106

Selected Bibliography

The collection and analysis of volatile hydrocarbon air pollutants using a timed elution chromatographic technique linked to a computer controlled mass spectrometer, Perry, R., Twibell, J. D., *Biomed. Mass Spectrom.*, **1**(1), 73–77 (1974).

Use of field ionization for analysis of complex hydrocarbon mixtures, Kuras, M., *Sb. Vys. Sk. Chem.-Technol. Praze, Technol. Paliv.*, **D33**, 189–213 (1976).

Characterization of volatile hydrocarbons in flowing seawater suspensions of Number 2 fuel oil, Bean, R. M., Blaylock, J. W., *Fate Eff. Pet. Hydrocarbons Mar. Ecosyst. Org., Proc. Symp.*, Pergamon, Elmsford, New York (1977), pp. 397–403.

The determination of volatile organic compounds in city air by gas chromatography combined with standard addition, selective subtraction, infrared spectrometry and mass spectrometry, Louw, C. W., Richards, J. F., Faure, P. K., *Atmos. Environ.*, **11**(8), 703–717 (1977).

Concentration and analysis of trace impurities in styrene monomer, Zlatkis, A., Anderson, J. W., Holzer, G., *J. Chromatogr.*, **142**, 127–129 (1977).

FLUORANTHENE

This compound, as for acenaphthene (*q.v.*), is not a polynuclear aromatic hydrocarbon, but comments on these compounds are pertinent.

FLUORANTHENE $C_{16}H_{10}$ (202)

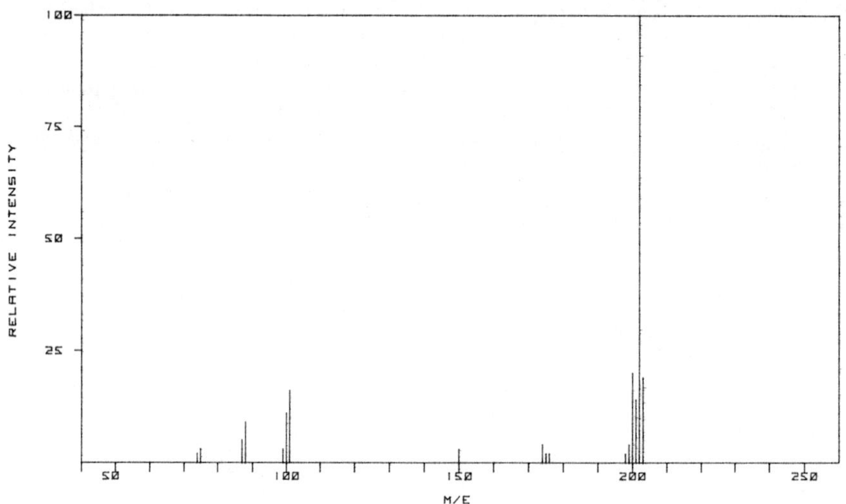

Spectral Data

Mass	Abundance	Mass	Abundance	Mass	Abundance
74	2.4	101	15.9	199	3.7
75	3.0	150	2.8	200	20.2
87	5.2	174	3.9	201	13.9
88	8.7	175	2.2	202	100.0
99	3.4	176	2.2	203	18.8
100	11.3	198	2.4		

Base–neutral extractable
CAS Name: Fluoranthene
Synonyms

Benzo[*jk*]fluorene
Idryl
1,2-(1,8-Naphthalenediyl)benzene
1,2-(1,8-Naphthylene)benzene

CAS Registry No.: 206-44-0
ROTECS Ref.: LL40250

Major Ions: 202, 200, 203, 101, 201, 100, 88, 87
EPA Ions: 202, 101, 100

Selected Bibliography

Analysis of complex polycyclic aromatic hydrocarbon mixtures by computerized GC/MS, Hites, R. A., *Prepr., Div. Pet. Chem., Am. Chem. Soc.*, **20**(4), 824–828 (1975).

Gas–liquid chromatographic evaluation and gas-chromatography/mass spectrometric application of new high-temperature liquid crystal stationary phases for polycyclic aromatic hydrocarbon separations, Janini, G. M., Muschik, G. M., Schroer, J. A., Zielinsk, W. L., Jr., *Anal. Chem.*, **48**(13), 1879–1883 (1976).

Gas chromatography–mass spectrometry of simulated arson residue using gasoline as an accelerant, Mach, M. H., *J. Forensic Sci.*, **22**(2), 348–357 (1977).

Determination of polynuclear aromatic hydrocarbons contaminated with chlorinated hydrocarbon pesticides, Negishi, T., *Bull. Environ. Contam. Toxicol.*, **19**(5), 545–548 (1978).

Determination of polycyclic aromatic hydrocarbons in atmospheric particulate matter by gas chromatography–mass spectrometry and high-pressure liquid chromatography, Thomas, R. S., Lao, R. C., Wang, D. T., Robinson, D., Sakuma, T., *Carcinogens–Comprehensive Survey*, vol. 3: *Polynuclear Aromatic Hydrocarbons:* Raven Press, New York (1978), pp. 9–19.

HALOETHERS

(other than those listed elsewhere)

Bis(2-chloroethoxy)methane, bis(2-chloroisopropyl) ether, 4-bromophenyl phenyl ether, and 4-chlorophenyl phenyl ether are the compounds included in this category. Haloethers listed elsewhere include bis(2-chloroethyl) ether, bis(2-chloromethyl) ether, and 2-chloroethyl vinyl ether (all listed under "chloroalkylethers").

BIS(2-CHLOROETHOXY)METHANE $C_5H_{10}Cl_2O_2$ (172)

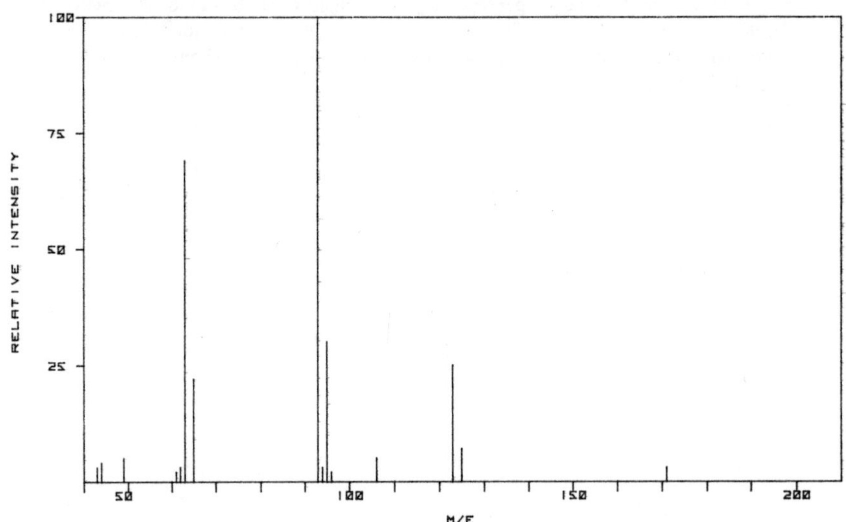

Spectral Data

Mass	Abundance	Mass	Abundance	Mass	Abundance
43	2.8	63	68.7	96	2.1
44	3.8	65	22.1	106	4.9
49	4.5	93	100.0	123	24.9
61	2.1	94	3.4	125	7.2
62	3.2	95	30.4	171	3.0

Base–neutral extractable
CAS Name: Methane, bis(2-chloroethoxy)-
Synonyms

Bis(β-chloroethyl)formal	Di-2-chloroethyl formal
Bis(2-chloroethyl) formal	Formaldehyde bis(β-chloroethyl) acetal
Dichloroethyl formal	Formaldehyde bis(2-chloroethyl) acetal

CAS Registry No.: 111-91-1
ROTECS Ref.: PA36750

Major Ions: 93, 63, 95, 123, 65, 125, 106, 49
EPA Ions: 93, 95, 123

BIS(2-CHLOROISOPROPYL) ETHER $C_6H_{12}Cl_2O$ (170)

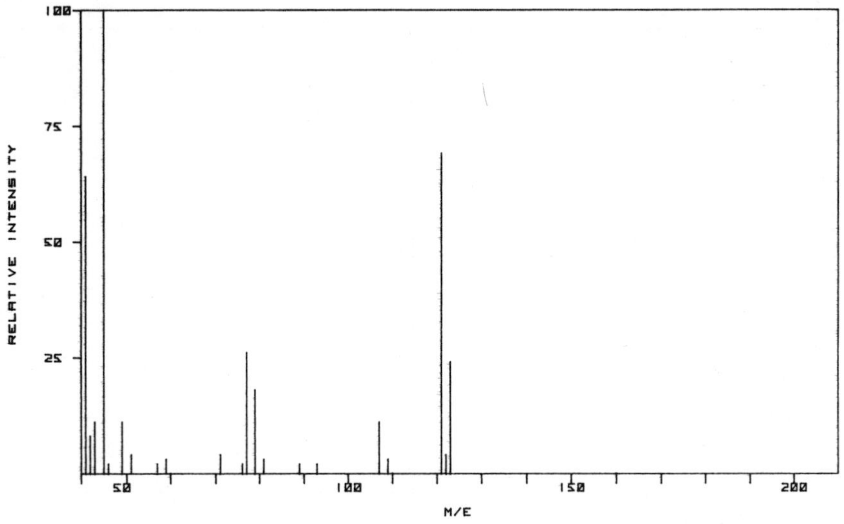

Spectral Data

Mass	Abundance	Mass	Abundance	Mass	Abundance
41	64.1	57	2.1	89	2.1
42	8.3	59	2.5	93	2.4
43	10.5	71	3.5	107	10.7
45	100.0	76	2.2	109	3.0
46	2.0	77	26.4	121	69.1
49	11.1	79	18.3	122	3.7
51	3.5	81	3.2	123	24.0

BIS(2-CHLOROISOPROPYL) ETHER–*continued*

Base–neutral extractable
CAS Name: Propane, 2,2′-oxybis(2-chloro)-
Synonyms

Bis(2-chloro-1-methylethyl) ether	2,2′-Dichloroisopropyl ether
Dichlorodiisopropyl ether	NCI-C50044
Dichloroisopropyl ether	2,2′-Oxybis(2-chloro)propane

CAS Registry No.: 39638-32-9
ROTECS Ref.: KN17500

Major Ions: 45, 121, 41, 77, 123, 79, 49, 107
EPA Ions: 45, 77, 79

4-BROMOPHENYL PHENYL ETHER $C_{12}H_9BrO$ (248)

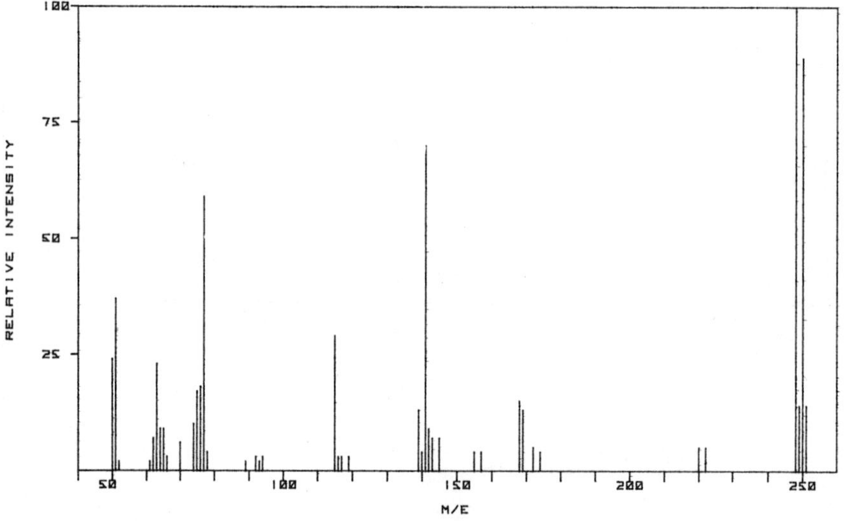

Spectral Data

Mass	Abundance	Mass	Abundance	Mass	Abundance
50	24.4	66	2.9	92	3.3
51	37.3	70	5.5	93	2.1
52	2.4	74	10.1	94	3.3
61	2.1	75	16.8	115	29.3
62	6.5	76	17.6	116	3.2
63	22.5	77	58.5	117	3.2
64	9.2	78	4.3	119	2.7
65	9.2	89	2.1	139	13.1

Spectral Data–*continued*

Mass	Abundance	Mass	Abundance	Mass	Abundance
140	3.9	157	4.0	222	4.6
141	69.5	168	14.6	248	100.0
142	8.7	169	12.5	249	14.0
143	7.4	172	4.5	250	88.6
145	6.7	174	4.3	251	13.5
155	4.0	220	4.8		

Base–neutral extractable
CAS Name: Ether, *p*-bromophenyl phenyl
Synonyms

- *p*-Bromodiphenyl ether
- 4-Bromodiphenyl ether
- *p*-Bromophenoxybenzene
- 4-Bromophenoxybenzene
- 1-Bromo-4-phenoxybenzene
- *p*-Bromophenyl ether
- 4-Bromophenyl ether
- *p*-Bromophenyl phenyl ether
- *p*-Phenoxybromobenzene

CAS Registry No.: 101-55-3

Major Ions: 248, 250, 141, 77, 51, 115, 50, 63
EPA Ions: 248, 250, 141

Selected Bibliography

Ion kinetic energy spectra of some chlorinated insecticides, Safe, S., Hutzinger, O., Jamieson, W. D., Cook, M., *Org. Mass Spectrom.*, 7(2), 217–224 (1973).

4-CHLOROPHENYL PHENYL ETHER $C_{12}H_9ClO$ (204)

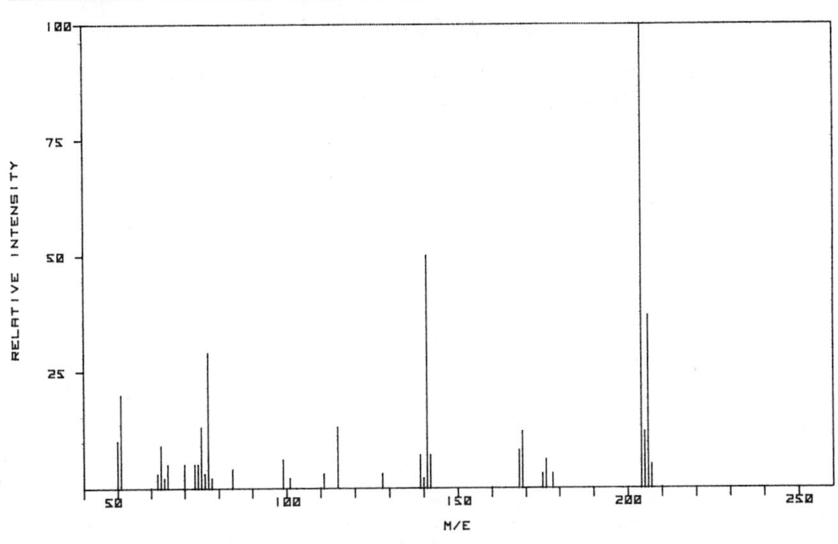

4-CHLOROPHENYL PHENYL ETHER–*continued*

Spectral Data

Mass	Abundance	Mass	Abundance	Mass	Abundance
50	10.3	77	28.7	142	7.2
51	20.3	78	2.2	168	7.5
62	3.2	84	3.8	169	11.5
63	9.0	99	6.1	175	2.7
64	2.4	101	2.2	176	6.4
65	5.3	111	3.2	178	2.6
70	5.1	115	12.7	204	100.0
73	4.8	128	3.4	205	12.0
74	5.3	139	7.0	206	36.9
75	12.7	140	2.2	207	4.7
76	2.9	141	50.0		

Base–neutral extractable
CAS Name: Ether, *p*-chlorophenyl phenyl
Synonyms
- 4-Chlorodiphenyl ether
- *p*-Chlorodiphenyl oxide
- *p*-Chlorophenyl phenyl ether
- 1-Chloro-4-phenoxybenzene

CAS Registry No.: 7005-72-3

Major Ions: 204, 141, 206, 77, 51, 115, 75, 205
EPA Ions: 204, 206, 141

Selected Bibliography

Ortho-effect in aromatic ethers, sulfides, and sulfoxides under electron impact, Granoth, I., *J. Chem. Soc., Perkin Trans. 2* (11), 1503–1505 (1972).

HALOMETHANES

(other than those listed elsewhere)

Bromoform, dibromochloromethane, bromodichloromethane, dichlorodifluoromethane, bromomethane, chloromethane, methylene chloride, and trichlorofluoromethane are the compounds listed in this category. Carbon tetrachloride and chloroform are listed separately.

Bromoform is used as a sedative and antitussive, but abuse leads to habituation or addiction.

Dichlorodifluoromethane is used as an aerosol propellant and a refrigerant.

Bromomethane is a soil fumigant and an insect fumigant used in freight cars, mills, ships, vaults, and warehouses. It is also used for degreasing wool and for extracting oils from flowers, nuts, and seeds. A minor application is in ionization chambers. This compound was formerly used in fire extinguishers.

Chloromethane is used as a refrigerant and was formerly employed as a local anesthetic.

Methylene chloride is a fruit ripening agent and a solvent for cellulose acetate. It is used in degreasing and cleaning fluids, and was formerly employed as an anesthetic.

Trichlorofluoromethane is used as an aerosol propellant and a refrigerant.

Several of the halomethanes, especially the trihalomethanes, are produced during the chlorination of water.

BROMOFORM $CHBr_3$ (250)

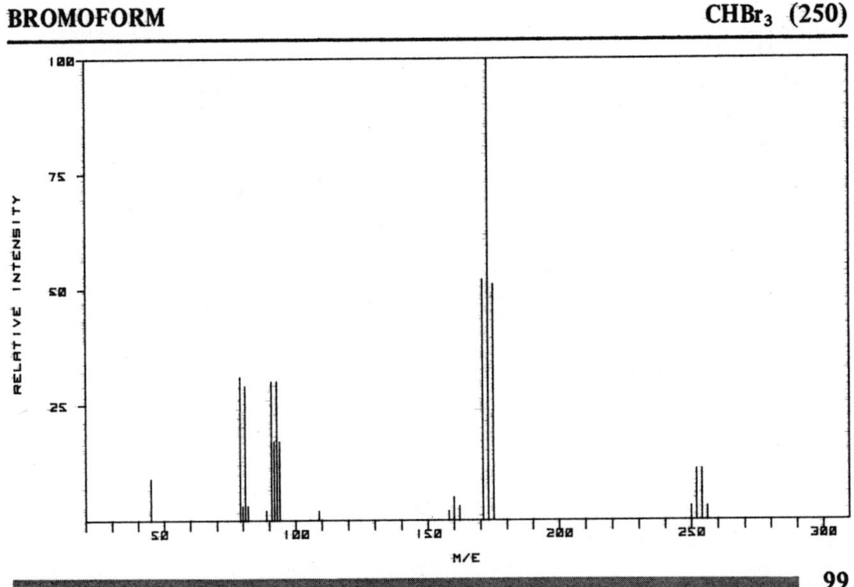

BROMOFORM–*continued*

Spectral Data

Mass	Abundance	Mass	Abundance	Mass	Abundance
45	9.0	92	16.8	171	51.9
79	30.6	93	29.6	173	100.0
80	2.7	94	16.9	175	50.7
81	28.9	109	2.3	250	3.3
82	2.9	158	2.3	252	10.8
89	2.0	160	5.0	254	10.8
91	29.7	162	2.5	256	3.3

Volatile
CAS Name: Methane, tribromo-
Synonyms
  Methenyl tribromide
  Tribromomethane

CAS Registry No.: 75-25-2
ROTECS Ref.: PB56000
Merck Index Ref.: 1418

Major Ions: 173, 171, 175, 79, 91, 93, 81, 94
EPA Ions: 171, 173, 175, 250, 252, 254, 256

Selected Bibliography

Volatile halogen compounds in the alga *Asparagopsis taxiformis* (Rhodophyta), Burreson, B. J., Moore, R. E., Roller, P. P., *J. Agric. Food Chem.*, **24**(4), 856–861 (1976).

Mass spectrometric determination of thermochemical data of tribromomethane and tetrabromomethane by study of their electron impact and heterogeneous pyrolytic decompositions, Kaposi, O., Riedel, M., Vass-Balthazar, K., Sanchez, G. R., Lelik, L., *Acta Chim. Acad. Sci. Hung.*, 89(3), 221–244 (1976).

Determination of the thermochemical data of tribromomethane and tetrabromomethane by mass spectrometry, investigating the electron-impact and heterogeneous pyrolytic decomposition, Kaposi, O., Riedel, M., Balthazar-Vass, K., Sanchez, R. G., Lelik, L., *Magy. Kem. Foly.*, **82**(4), 155–166 (1976).

Direct aqueous injection gas chromatography–mass spectrometry for analysis of organohalides in water at concentrations below the parts per billion level, Fujii, T., *J. Chromatogr.*, **139**(2), 297–302 (1977).

Mass spectrometric identification of radicals originating from heterogeneous pyrolytic decomposition of bromomethanes, Lelik, L., Keszei, E., *Magy. Kem. Lapja*, **32**(11), 619–624 (1977).

DIBROMOCHLOROMETHANE $CHBr_2Cl$ (206)

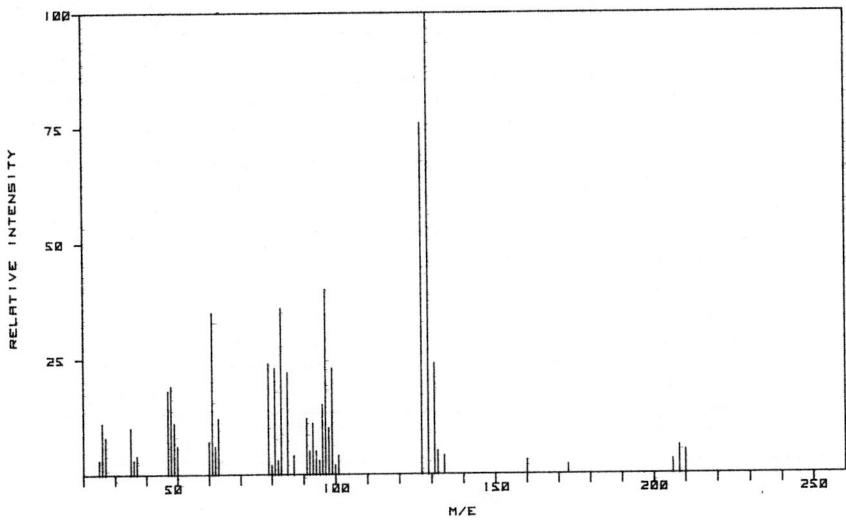

Spectral Data

Mass	Abundance	Mass	Abundance	Mass	Abundance
25	3.2	79	24.1	98	10.4
26	10.6	80	2.1	99	22.8
27	7.5	81	23.4	100	2.2
35	10.1	82	2.6	101	3.9
36	3.4	83	36.0	127	76.2
37	4.1	85	22.2	129	100.0
47	18.4	87	3.6	131	23.7
48	18.5	91	11.8	132	4.6
49	11.0	92	5.1	134	4.3
50	5.9	93	11.1	160	2.6
60	7.1	94	5.4	173	2.2
61	35.2	95	2.6	206	2.6
62	6.3	96	15.1	208	6.2
63	11.8	97	39.6	210	4.6

Volatile
CAS Name: Methane, dibromochloro-
Synonym
Chlorodibromomethane

CAS Registry No.: 124-48-1

Major Ions: 129, 127, 97, 83, 61, 79, 131, 81
EPA Ions: 129, 127, 208, 206

DIBROMOCHLOROMETHANE–*continued*

Selected Bibliography

Volatile halogen compounds in the alga *Asparagopsis taxiformis* (Rhodophyta), Burreson, B. J., Moore, R. E., Roller, P. P., *J. Agric. Food Chem.*, **24**(4), 856–861 (1976).

The determination of traces of organohalogen compounds in aqueous solution by direct injection gas chromatography–mass spectrometry and single ion detection, Fujii, T., *Anal. Chim. Acta*, **92**(1), 117–122 (1977).

Direct aqueous injection gas chromatography-mass spectrometry for analysis of organohalides in water at concentrations below the parts per billion level, Fujii, T., *J. Chromatogr.*, **139**(2), 297–302 (1977).

BROMODICHLOROMETHANE $CHBrCl_2$ (162)

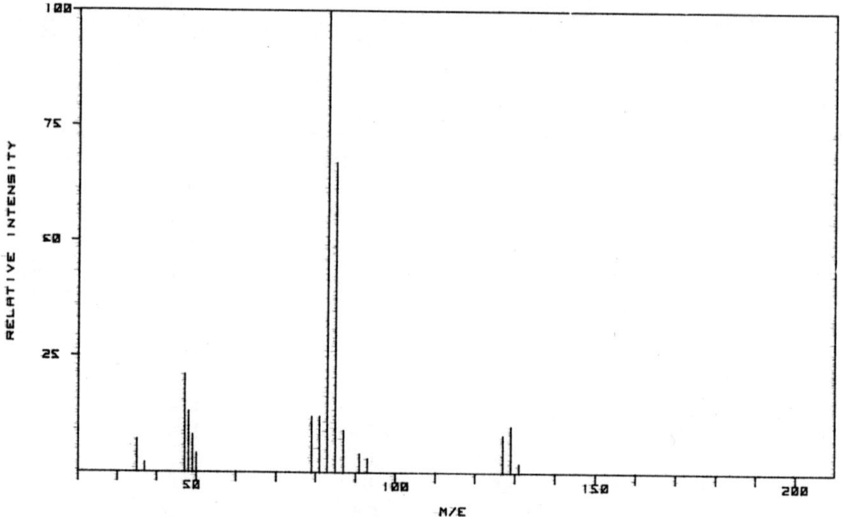

Spectral Data

Mass	Abundance	Mass	Abundance	Mass	Abundance
35	6.8	79	12.4	91	3.5
37	2.2	81	11.9	93	3.4
47	20.8	83	100.0	127	7.5
48	12.8	85	66.5	129	10.1
49	7.8	87	9.4	131	2.2
50	4.1				

Volatile
CAS Name: Methane, bromodichloro-
Synonym
Dichlorobromomethane

CAS Registry No.: 75-27-4

Major Ions: 83, 85, 47, 48, 79, 81, 129, 87
EPA Ions: 83, 85, 127, 129

Selected Bibliography

The determination of traces of organohalogen compounds in aqueous solution by direct injection gas chromatography–mass spectrometry and single ion detection, Fujii, T., *Anal. Chim. Acta*, **92**(1), 117–122 (1977)

Direct aqueous injection gas chromatography–mass spectrometry for analysis of organohalides in water at concentrations below the parts per billion level, Fujii, T., *J. Chromatogr.*, **139**(2), 297–302 (1977).

DICHLORODIFLUOROMETHANE CCl_2F_2 (120)

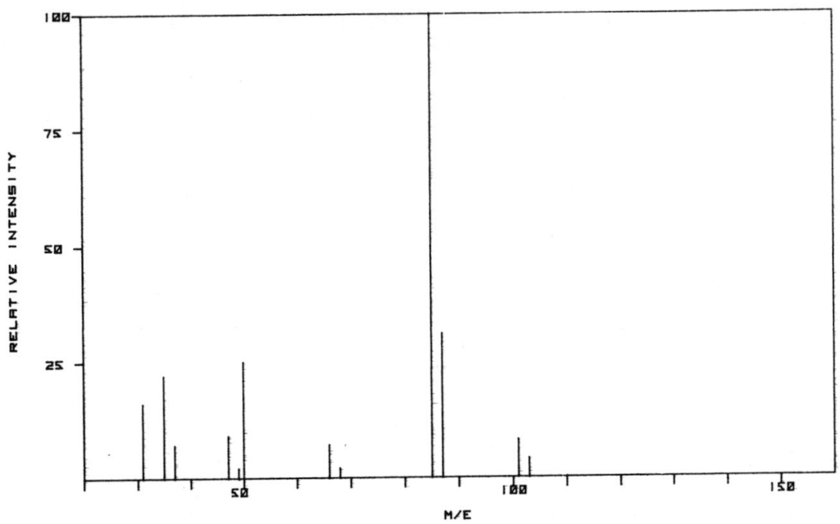

Spectral Data

Mass	Abundance	Mass	Abundance	Mass	Abundance
31	16.3	49	2.2	85	100.0
35	22.0	50	24.6	87	30.9
37	6.7	66	6.8	101	7.6
47	8.7	68	2.1	103	4.3

DICHLORODIFLUOROMETHANE–*continued*

Volatile
CAS Name: Methane, dichlorodifluoro-
Synonyms

Algofrene type 2	Freon 12	Isotron 12
Arcton 6	Freon F-12	Ledon 12
Difluorodichloromethane	Frigen 12	Propellant 12
Electro-CF 12	Genetron 12	R 12
Eskimon 12	Halon	Refrigerant 12
FC 12	Isceon 122	Ucon 12
Fluorocarbon 12		

CAS Registry No.: 75-71-8
ROTECS Ref.: PA82000
Merck Index Ref.: 3038

Major Ions: 85, 87, 50, 35, 31, 47, 101, 66
EPA Ions: 85, 87, 101, 103

Selected Bibliography

Mass spectrometric study of fluorochloro-substituted ethylenes, Syrvatka, B. G., Gil'burd, M. M., Bel'ferman, A. L., *Zh. Org. Khim.*, 8(8), 1553–1557 (1972).

Mass spectral intensities of inorganic fluorine-containing compounds, Beattie, W. H., *Appl. Spectrosc.*, 29(4), 334–337 (1975).

Multiple pollutant monitoring using spectroscopic gas-chromatographic methods in a mobile laboratory, Hollingdale-Smith, P. A., *Proc. Anal. Div. Chem. Soc.*, 12(12), 317–319 (1975).

Chlorinated hydrocarbons in the atmosphere. Analysis at the parts-per-trillion level by GC–MS (gas chromatography–mass spectrometry), Tyson, B. J., *Anal. Lett.*, 8(11), 807–813 (1975).

Terminal ions in weak atmospheric pressure plasmas. Applications of atmospheric pressure ionization to trace impurity analysis in gases, Siegel, M. W., Fite, W. L., *J. Phys. Chem*, 80(26), 2871–2881 (1976).

BROMOMETHANE CH_3Br (94)

Spectral Data

Mass	Abundance	Mass	Abundance	Mass	Abundance
79	21.3	91	9.0	94	100.0
80	4.3	92	4.3	95	14.1
81	20.5	93	21.6	96	89.3
82	3.9				

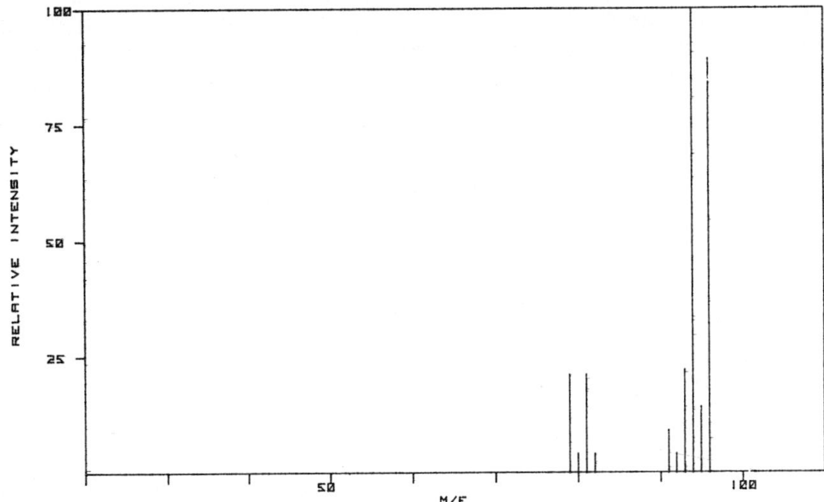

Volatile
CAS Name: Methane, bromo-
Synonyms

Brom-o-gas	Halon 1001	Metabrom	Pestmaster
Celfume	Haltox	Metafume	Profume
Dowfume MC-2	Iscobrome	Methogas	Rotox
Dowfume MC-33	MB	Methylbromid	Terabol
EDCO	MBX	Methyl bromide	Terr-o-gas 100
Embafume	MEBR	Monobromomethane	Zytox

CAS Registry No.: 74-83-9
ROTECS Ref.: PA49000
Merck Index Ref.: 5904

Major Ions: 94, 96, 93, 79, 81, 95, 91, 92
EPA Ions: 94, 96

Selected Bibliography

Fragmentation mechanisms in methyl X (X = amine, hydroxide, hydrosulfide, chloride, bromide, and iodide) interpreted by the molecular orbital method, Ikuta, S., Yoshihara, K., Shiokawa, T., *Shitsuryo Bunseki*, 22(4), 233–238 (1974).

Mass spectrometric investigations on the electron impact and heterogeneous pyrolytic decomposition of methyl bromide, Kaposi, O., Riedel, M., Sanchez, G. R., *Magy. Kem. Foly.*, 80(9), 419–428 (1974).

Mass spectrometric study of electron impact and heterogeneous pyrolytic decomposition of methyl bromide, Kaposi, O., Riedel, M., Sanchez, G. R., *Acta Chim. Acad. Sci. Hung.*, 85(4), 361–382 (1975).

Unimolecular dissociations and internal conversions of methyl halide ions, Eland, J. H. D., Frey, R., Kuestler, A., Schulte, H., Brehm, B., *Int. J. Mass Spectrom. Ion Phys.*, 22(1–2), 155–170 (1976).

Capacitive integration to produce high precision isotope ratio measurements on methyl chloride and methyl bromide samples, Willey, J. F., Taylor, J. W., *Anal. Chem.*, 50(13), 1930–1933 (1978).

CHLOROMETHANE CH_3Cl (50)

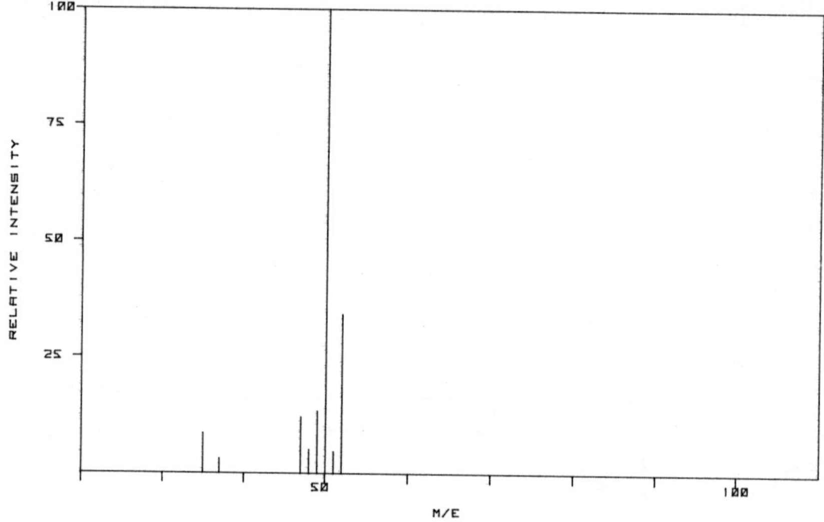

Spectral Data

Mass	Abundance	Mass	Abundance	Mass	Abundance
35	8.4	48	5.0	51	4.6
37	3.0	49	13.3	52	34.2
47	11.9	50	100.0		

Volatile
CAS Name: Methane, chloro-
Synonyms
- Artic
- Freon 40
- Methyl chloride
- Monochloromethane

CAS Registry No.: 74-87-3
ROTECS Ref.: PA63000
Merck Index Ref.: 5916

Major Ions: 50, 52, 49, 47, 35, 48, 51, 37
EPA Ions: 50, 52

Selected Bibliography

Fragmentation mechanisms in methyl X (X = amine, hydroxide, hydrosulfide, chloride, bromide, and iodide) interpreted by the molecular orbital method, Ikuta, S., Yoshihara, K., Shiokawa, T., *Shitsuryo Bunseki*, **22**(4), 233–238 (1974).

Photoionization study of the ionization potentials and fragmentation paths of the chlorinated methanes and carbon tetrabromide, Werner, A. S., Tsai, B. P., Baer, T., *J. Chem. Phys.*, **60**(9), 3650–3657 (1974).

Unimolecular dissociations and internal conversions of methyl halide ions, Eland, J. H. D., Frey, R., Kuestler, A., Schulte, H., Brehm, B., *Int. J. Mass Spectrom. Ion Phys.*, **22**(1–2), 155–170 (1976).

Kinetic energies of fragment ions from some hydrocarbons and organic halides in a modified mass spectrometer, Ossinger, A. I., Weiner, E. R., *J. Chem. Phys.*, **65**(7), 2892–2900 (1976).

Capacitive integration to produce high precision isotope ratio measurements on methyl chloride and methyl bromide samples, Willey, J. F., Taylor, J. W., *Anal. Chem.*, **50**(13), 1930–1933 (1978).

METHYLENE CHLORIDE CH_2Cl_2 (84)

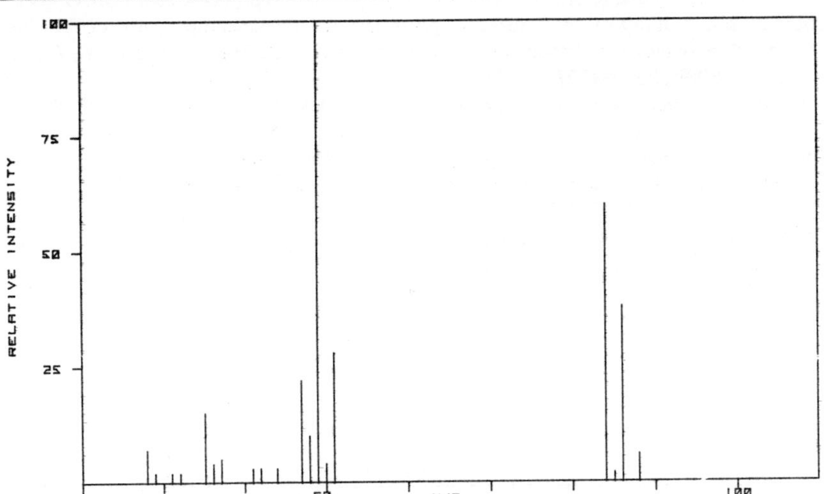

Spectral Data

Mass	Abundance	Mass	Abundance	Mass	Abundance
28	6.6	41	3.2	50	4.3
29	2.0	42	2.5	51	28.4
31	2.1	44	2.8	84	60.1
32	2.2	47	22.0	85	2.3
35	14.6	48	9.6	86	38.3
36	4.2	49	100.0	88	5.8
37	5.4				

METHYLENE CHLORIDE–*continued*

Volatile
CAS Name: Methane, dichloro-
Synonyms

Aerothene MM	Methylene bichloride	NCI-C50102
Dichloromethane	Methylene dichloride	Solaesthin
Freon 30	Narkotil	Solmethine
Methane dichloride		

CAS Registry No.: 75-09-2
ROTECS Ref.: PA80500
Merck Index Ref.: 5932

Major Ions: 49, 84, 86, 51, 47, 35, 48, 28
EPA Ions: 49, 51, 84, 86

Selected Bibliography

Mass spectrometric analysis of product water from coal gasification, Schmidt, C. E., Sharkey, A. G., Jr., Friedel, R. A., *U.S. Bur. Mines, Tech. Prog. Rep.*, (TPR 86), 7 pp. (1974).

Photoionization study of the ionization potentials and fragmentation paths of the chlorinated methanes and carbon tetrabromide, Werner, A. S., Tsai, B. P., Baer, T., *J. Chem. Phys.*, **60**(9), 3650–3657 (1974).

Volatile flavor components of leek, Schreyen, L., Dirinck, P., Van Wassenhove, F., Schamp, N., *J. Agric. Food Chem.*, **24**(2), 336–341 (1976).

Direct aqueous injection gas chromatography–mass spectrometry for analysis of organohalides in water at concentrations below the parts per billion level, Fujii, T., *J. Chromatogr.*, **139**(2), 297–302 (1977).

Concentration and analysis of trace impurities in styrene monomer, Zlatkis, A., Anderson, J. W., Holzer, G., *J. Chromatogr.*, **142**, 127–129 (1977).

TRICHLOROFLUOROMETHANE CCl_3F (136)

Spectral Data

Mass	Abundance	Mass	Abundance	Mass	Abundance
28	2.2	49	3.9	84	2.5
31	9.3	66	17.6	101	100.0
35	12.8	68	6.0	103	59.1
37	4.5	82	4.1	105	11.0
47	10.9				

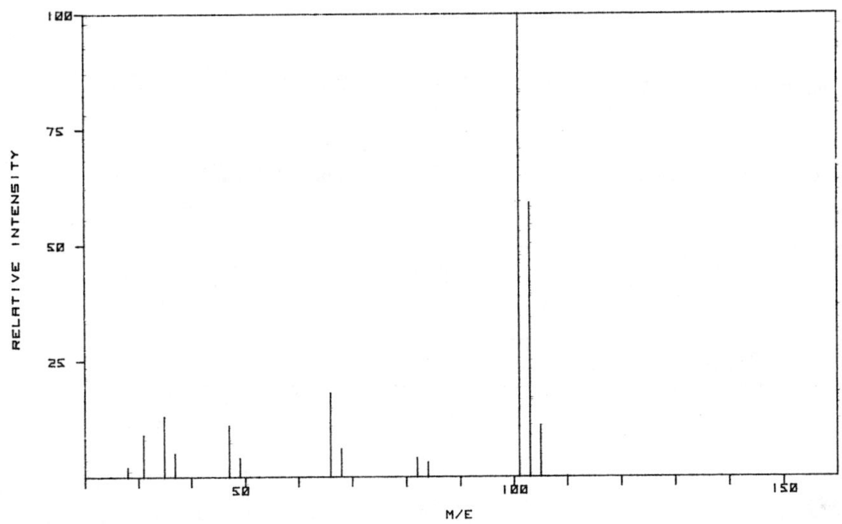

Volatile

CAS Name: Methane, trichlorofluoro-

Synonyms

Algofrene type 1	Freon 11	Ledon 11
Arcton 9	Freon MF	Monofluorotrichloromethane
Electro-CF 11	Freon R 11	NCI-C04637
Eskimon 11	Frigen 11	Propellant 11
F 11B	Frigen 11A	R 11
FC 11	Genetron 11	Trichloromonofluoromethane
FKW 11	Isceon 131	Ucon fluorocarbon 11
Fluorochloroform	Isotron 11	Ucon refrigerant 11
Fluorotrichloromethane		

CAS Registry No.: 75-69-4
ROTECS Ref.: PB61250
Merck Index Ref.: 9320

Major Ions: 101, 103, 66, 35, 105, 47, 31, 68
EPA Ions: 101, 103

Selected Bibliography

Chlorinated hydrocarbons in the atmosphere. Analysis at the parts-per-trillion level by GC-MS (gas chromatography-mass spectrometry), Tyson, B. J., *Anal. Lett.*, 8(11), 807-813 (1975).

Terminal ions in weak atmospheric pressure plasmas. Applications of atmospheric pressure ionization to trace impurity analysis in gases, Siegel, M. W., Fite, W. L., *J. Phys. Chem.*, 80(26), 2871–2881 (1976).

HEPTACHLOR AND METABOLITES

Heptachlor and heptachlor epoxide are the compounds included in this category.

Heptachlor is a powerful, but persistent, contact insecticide. Its use is now severely restricted in the U.S.

Heptachlor epoxide is formed by autoxidation and metabolism.

HEPTACHLOR $C_{10}H_5Cl_7$ (370)

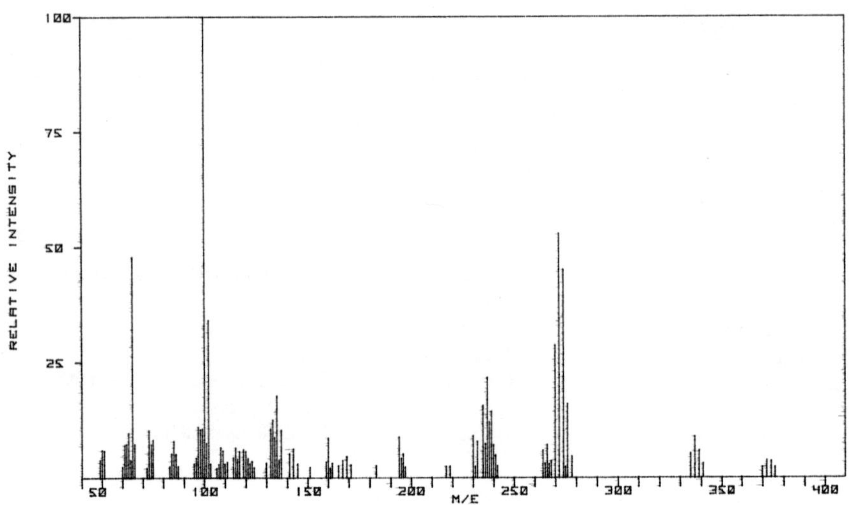

Spectral Data

Mass	Abundance	Mass	Abundance	Mass	Abundance
49	3.9	75	8.2	102	34.1
50	6.0	83	2.4	103	3.0
51	5.9	84	5.2	106	2.1
60	2.4	85	7.9	107	2.9
61	7.1	86	5.1	108	6.6
62	7.3	87	2.5	109	5.9
63	9.7	95	3.0	110	3.0
64	3.8	96	4.3	111	3.4
65	47.8	97	11.0	114	4.4
66	7.3	98	10.4	115	6.5
72	2.2	99	10.5	116	4.1
73	10.3	100	100.0	117	5.7
74	7.2	101	7.4	119	6.1

Spectral Data–*continued*

Mass	Abundance	Mass	Abundance	Mass	Abundance
120	5.8	167	3.7	242	2.5
121	4.2	169	4.5	264	5.8
122	3.0	171	2.7	265	3.0
123	3.6	183	2.5	266	7.1
124	2.2	194	8.8	267	2.9
130	3.2	195	4.2	268	3.5
132	10.6	196	5.1	270	28.5
133	12.5	197	2.3	272	52.8
134	8.7	217	2.4	274	45.0
135	17.7	219	2.4	275	2.3
136	3.8	230	9.0	276	15.9
137	10.3	231	2.4	278	4.5
141	5.2	232	7.8	335	5.2
143	6.2	235	15.5	337	8.8
145	3.0	236	7.3	339	5.8
151	2.3	237	21.6	341	3.0
159	3.4	238	11.9	370	2.2
160	8.6	239	14.2	372	3.6
161	2.1	240	7.0	374	3.5
162	3.1	241	4.8	376	2.2
165	2.5				

Pesticide

CAS Name: 4,7-Methanoindene, 1,4,5,6,7,8,8-heptachloro-3a,4,7,7a-tetrahydro-

Synonyms

- Aahepta
- Agroceres
- 3-Chlorochlordene
- Drinox
- E 3314
- Ent 15,152
- GPKh
- H
- H-34
- Hepta
- Heptachlorane
- Heptachlorodicyclopentadiene
- 1,4,5,6,7,8,8-Heptachloro-3a,4,7,7a-tetrahydro-4,7-endomethanoindene
- 1,4,5,6,7,8,8-Heptachloro-3a,4,7,7a-tetrahydro-4,7-methanoindene
- 1(3a),4,5,6,7,8,8-Heptachloro-3a(1),4,-7,7a-tetrahydro-4,7-methanoindene
- 3a,4,5,6,7,8,8-Heptachloro-3a,4,7,7a-tetrahydro-4,7-methanoindene
- 1,4,5,6,7,8,8-Heptachloro-3a,4,7,7a-tetrahydro-4,7-methylene indene
- 1,4,5,6,7,10,10-Heptachloro-4,7,8,9-tetrahydro-4,7-endomethyleneindene
- Heptagran
- Heptalube
- Heptamul
- NCI-C00180
- Rhodiachlor
- Velsicol 104
- Velsicol heptachlor

CAS Registry No.: 76-44-8 (formerly 23720-59-4, 37229-06-4)
ROTECS Ref.: PC07000
Merck Index Ref.: 4514

Major Ions: 100, 272, 65, 274, 102, 270, 237, 135
EPA Ions: 100, 272, 274

HEPTACHLOR–*continued*

Selected Bibliography

Applications of mass spectrometry to trace determinations of environmental toxic materials, Abramson, F. P., *Anal. Chem.*, **44**(14), 28A–33A,35A (1972).

Positive chemical ionization mass spectra of polycyclic chlorinated pesticides, Biros, F. J., Dougherty, R. C., Dalton, J., *Org. Mass Spectrom.*, **6**(11), 1161–1169 (1972).

Gas chromatographic and direct inlet mass spectra of heptachlor and 1-hydroxychlordene, Demayo, A., Comba, M., *Bull. Environ. Contam. Toxicol.*, **8**(4), 212–216 (1972).

Negative chemical ionization mass spectra of polycyclic chorinated insecticides, Dougherty, R. C., Dalton, J., Biros, F. J., *Org. Mass Spectrom.*, **6**(11), 1171–1181 (1972).

Application of coupled gas chromatography–mass spectrometry in methods for the study and determination of pesticide residues and organic micropollutants in environmental and food materials, Mestres, R., Chevallier, C., Espinoza, C., Cornet, R., *Ann. Falsif. Exper. Chim.*, **70**(751), 177–188 (1977).

HEPTACHLOR EPOXIDE — $C_{10}H_5Cl_7O$ (386)

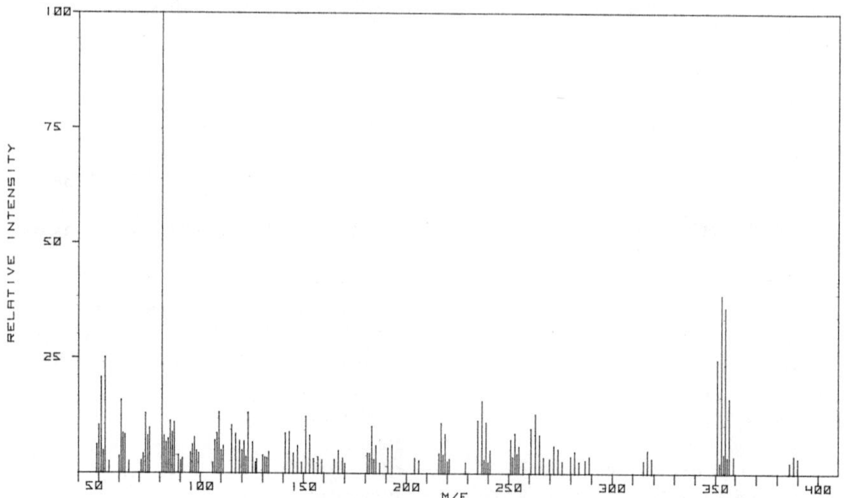

Spectral Data

Mass	Abundance	Mass	Abundance	Mass	Abundance
49	6.0	55	2.4	65	2.5
50	10.2	60	3.6	71	2.6
51	20.7	61	15.7	72	4.1
52	4.7	62	8.5	73	12.8
53	25.0	63	8.2	74	8.0

Spectral Data–*continued*

Mass	Abundance	Mass	Abundance	Mass	Abundance
75	9.7	141	8.5	240	2.3
81	100.0	143	8.8	241	4.9
82	8.0	145	4.2	251	7.2
83	6.5	147	5.8	252	3.5
84	7.3	149	2.3	253	8.5
85	11.2	151	12.2	254	4.1
86	8.7	153	8.1	255	5.7
87	10.9	155	3.0	257	2.3
89	3.8	157	3.5	261	9.6
90	2.6	159	2.8	263	12.8
91	3.1	165	2.9	265	8.2
95	4.3	167	4.9	267	3.3
96	6.0	169	3.2	270	3.0
97	7.7	170	2.0	272	5.8
98	4.7	181	4.3	274	5.2
99	4.2	182	4.2	276	2.5
106	2.2	183	10.0	280	3.6
107	7.0	184	2.9	282	4.6
108	8.5	185	5.9	284	2.4
109	13.2	187	2.2	287	2.7
110	4.9	191	5.4	289	3.6
111	5.8	193	6.0	315	2.6
115	10.2	204	3.2	317	4.9
117	8.4	206	2.7	319	3.1
119	6.9	216	4.2	351	24.6
120	4.9	217	10.8	352	2.2
121	6.8	218	3.9	353	38.6
122	3.5	219	8.4	354	4.0
123	13.1	220	2.5	355	35.9
125	6.6	221	3.0	356	3.3
127	2.9	229	2.3	357	16.2
130	3.8	235	11.3	359	3.6
131	3.3	237	15.6	386	2.3
132	3.2	238	2.8	388	3.8
133	4.5	239	10.9	390	3.1

Pesticide
CAS Name: 4,7-Methanoindan, 1,4,5,6,7,8,8-heptachloro-2,3-epoxy-3a,4,7,7a-tetrahydro-
Synonyms

- Ent 25,584
- Epoxyheptachlor
- HCE
- 1,4,5,6,7,8,8-Heptachloro-2,3-epoxy-2,3,3a,4,7,7a-hexahydro-4,7-methanoindene
- 1,4,5,6,7,8,8-Heptachloro-2,3-epoxy-3a,4,7,7a-tetrahydro-4,7-methanoindan
- Hiptachlor epoxide
- Velsicol 53-CS-17

HEPTACHLOR EPOXIDE–*continued*

CAS Registry No.: 1024-57-3 (formerly 4067-30-5, 23720-62-9, 24717-72-4)
ROTECS Ref.: PB94500

Major Ions: 81, 353, 355, 53, 351, 51, 357, 61
EPA Ions: 353, 355, 351

Selected Bibliography

Negative chemical ionization mass spectra of polycyclic chlorinated insecticides, Dougherty, R. C., Dalton, J., Biros, F. J., *Org. Mass Spectrom.*, **6**(11), 1171–1181 (1972).

Application of coupled gas chromatography–mass spectrometry in methods for the study and determination of pesticide residues and organic micropollutants in environmental and food materials, Mestres, R., Chevallier, C., Espinoza, C., Cornet, R., *Ann. Falsif. Exper. Chim.*, **70**(751), 177–188 (1977).

HEXACHLOROBUTADIENE

This compound is an intermediate in pesticide manufacture.

HEXACHLOROBUTADIENE C_4Cl_6 (258)

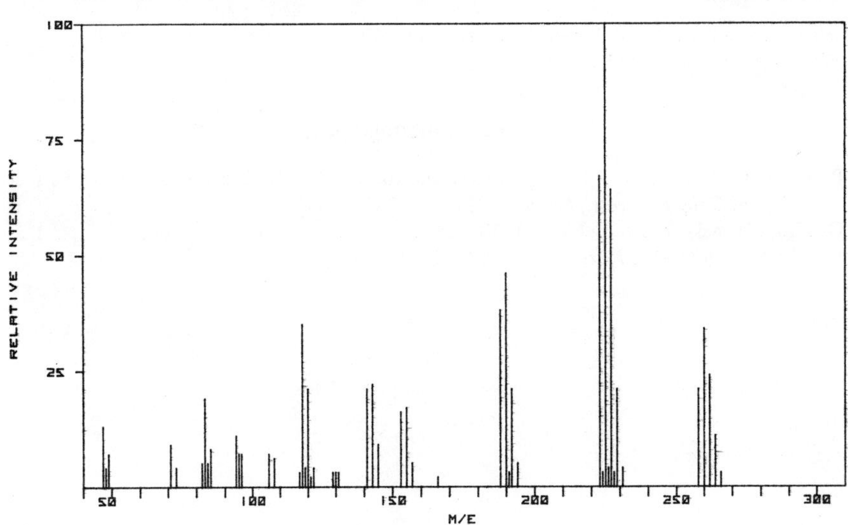

Spectral Data

Mass	Abundance	Mass	Abundance	Mass	Abundance
47	13.1	119	4.3	191	2.8
48	4.0	120	21.4	192	21.1
49	6.6	121	2.3	194	4.6
71	8.5	122	3.7	223	67.2
73	4.0	129	3.1	224	3.1
82	5.4	130	2.6	225	100.0
83	19.1	131	2.6	226	4.3
84	5.1	141	21.1	227	64.1
85	7.7	143	21.9	228	3.1
94	10.8	145	8.5	229	20.5
95	6.6	153	16.2	231	3.7
96	6.8	155	17.1	258	20.8
106	6.6	157	5.4	260	33.6
108	6.0	166	2.3	262	23.6
117	2.8	188	37.9	264	10.8
118	34.8	190	45.9	266	3.1

HEXACHLOROBUTADIENE–*continued*

Base–neutral extractable
CAS Name: 1,3-Butadiene, 1,1,2,3,4,4-hexachloro-
Synonyms

C-46	Hexachloro-1,3-butadiene
HCBD	1,1,2,3,4,4-Hexachloro-1,3-butadiene
Hexachlorbutadiene	Perchlorobutadiene

CAS Registry No.: 87-68-3
ROTECS Ref.: EJ07000

Major Ions: 225, 223, 227, 190, 188, 118, 260, 262
EPA Ions: 225, 223, 227

Selected Bibliography

Positive-ion mass spectra of some perhalogenated compounds, Contineanu, M. A., Grubel, K., *An. Univ. Bucuresti, Chim.*, **20**(2), 175–181 (1971).

Quadrupole mass spectrometer of 1–300 a.m.u. mass range, Berecz, I., Bohatka, S., Gal, J., Paal, A., *ATOMKI Kozl.*, **19**(2), 123–134 (1977).

HEXACHLOROCYCLOHEXANE

The α-, β-, γ-, and δ-isomers are the compounds included in this category.

The major components of the insecticide formulations are the α-, β-, and γ-isomers, of which the last is the most potent. Other isomers (ε, η, θ, and ζ) also exist. Pharmaceutical preparations contain 99% of the γ-isomer. This compound is used as a pediculicide and a scabicide and (in veterinary medicine) as an ectoparasiticide.

α-BHC **$C_6H_6Cl_6$ (288)**

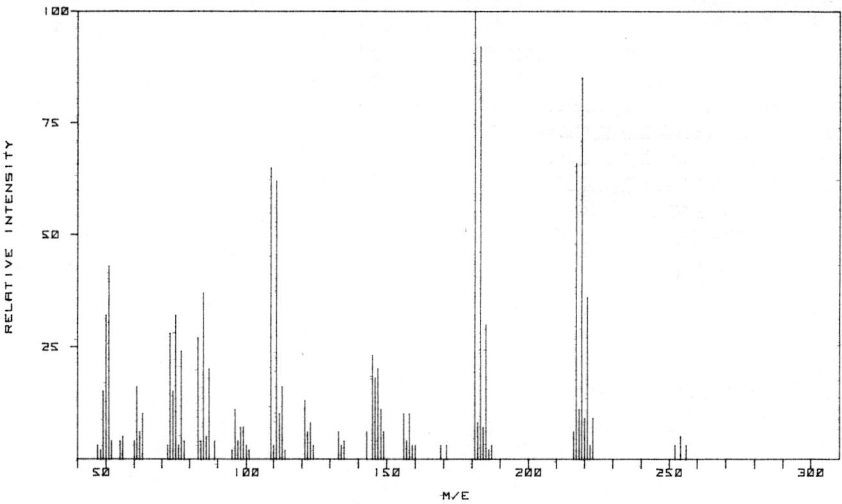

Spectral Data

Mass	Abundance	Mass	Abundance	Mass	Abundance
47	2.7	56	4.5	74	14.8
48	2.3	60	4.1	75	31.8
49	14.6	61	16.3	76	3.2
50	31.6	62	6.3	77	23.9
51	43.0	63	10.2	78	4.2
52	4.4	72	2.7	83	26.8
55	4.4	73	27.7	84	4.1

α-BHC–*continued*

Spectral Data–*continued*

Mass	Abundance	Mass	Abundance	Mass	Abundance
85	36.9	123	7.9	181	100.0
86	4.9	124	2.9	182	8.1
87	19.8	133	5.8	183	91.7
89	4.0	134	2.6	184	6.7
95	2.4	135	4.4	185	30.3
96	11.4	143	5.9	186	2.1
97	4.0	145	23.4	187	3.1
98	7.3	146	17.8	216	5.5
99	6.8	147	19.5	217	66.0
100	2.7	148	10.9	218	10.6
101	2.0	149	5.7	219	84.7
109	65.3	156	10.1	220	9.2
110	3.2	157	3.5	221	35.7
111	62.4	158	9.6	222	3.3
112	9.8	159	2.6	223	9.0
113	15.8	160	2.9	252	2.6
114	2.5	169	2.8	254	5.3
121	12.9	171	2.9	256	3.3
122	6.0				

Pesticide
CAS Name: Cyclohexane, 1,2,3,4,5,6-hexachloro-, α-
Synonyms

- Benzene hexachloride, α-isomer
- α-Benzenehexachloride
- Ent 9,232
- α-HCH
- α-Hexachloran
- α-Hexachlorane
- α-Hexachlorocyclohexane
- α-1,2,3,4,5,6-Hexachlorocyclohexane
- α-Lindane

CAS Registry No.: 319-84-6 (formerly 20437-97-2)
ROTECS Ref. No.: GV35000

Major Ions: 181, 183, 219, 217, 109, 111, 51, 85
EPA Ions: 183, 109, 181

Selected Bibliography

Ion kinetic energy spectra of some chlorinated insecticides, Safe, S., Hutzinger, O., Jamieson, W. D., Cook, M., *Org. Mass Spectrom.*, 7(2), 217–224 (1973).

β-BHC $C_6H_6Cl_6$ (288)

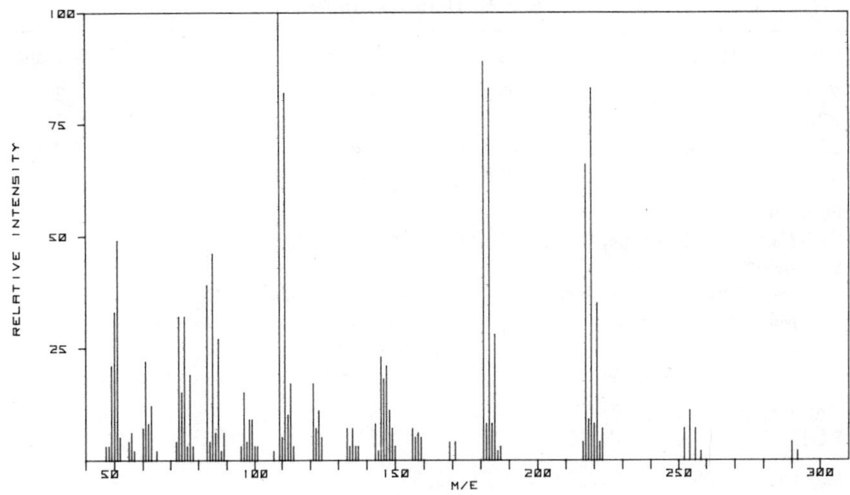

Spectral Data

Mass	Abundance	Mass	Abundance	Mass	Abundance
47	3.1	87	26.6	137	2.9
48	2.7	88	2.4	143	7.9
49	21.0	89	5.6	144	2.0
50	33.4	95	3.3	145	22.5
51	48.8	96	15.0	146	18.0
52	4.7	97	4.1	147	20.5
55	3.6	98	9.2	148	11.4
56	5.5	99	9.3	149	7.4
57	2.0	100	2.9	150	2.5
60	6.8	101	3.0	156	6.6
61	22.4	107	2.2	157	5.0
62	8.3	109	100.0	158	5.8
63	12.4	110	5.2	159	5.0
65	2.0	111	82.4	169	3.6
72	3.8	112	9.9	171	3.5
73	31.9	113	17.1	181	89.3
74	14.6	114	2.6	182	7.7
75	32.2	121	17.1	183	82.7
76	2.9	122	6.6	184	7.7
77	18.5	123	11.1	185	27.7
78	3.3	124	5.0	186	2.4
83	38.6	133	6.8	187	2.9
84	3.9	134	3.1	216	3.8
85	46.0	135	7.4	217	65.6
86	6.2	136	2.9	218	8.6

β-BHC–*continued*

Spectral Data–*continued*

Mass	Abundance	Mass	Abundance	Mass	Abundance
219	82.9	223	7.4	258	2.2
220	8.2	252	7.2	290	3.5
221	34.7	254	10.8	292	2.0
222	4.0	256	6.7		

Pesticide

CAS Name: Cyclohexane, 1,2,3,4,5,6-hexachloro-, β-

Synonyms

- *trans*-β-Benzenehexachloride
- β-Benzenehexachloride
- β-HCH
- β-Hexachlorobenzene
- β-Hexachlorocyclohexane
- β-1,2,3,4,5,6-Hexachlorocyclohexane
- β-Lindane

CAS Registry No.: 319-85-7
ROTECS Ref. No.: GV43750

Major Ions: 109, 181, 183, 111, 217, 51, 85, 83
EPA Ions: 181, 183, 109

Selected Bibliography

Ion kinetic energy spectra of some chlorinated insecticides, Safe, S., Hutzinger, O., Jamieson, W. D., Cook, M., *Org. Mass Spectrom.*, 7(2), 217–224 (1973).

γ-BHC $C_6H_6Cl_6$ (288)

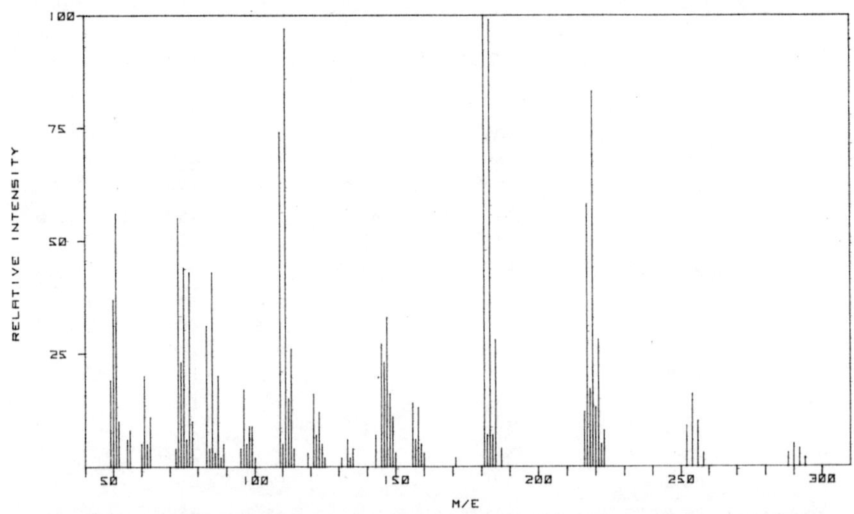

HEXACHLOROCYCLOHEXANE

Spectral Data

Mass	Abundance	Mass	Abundance	Mass	Abundance
49	19.3	98	9.2	157	6.3
50	36.9	99	8.8	158	13.4
51	55.6	100	2.3	159	4.6
52	9.5	109	74.4	160	3.3
55	5.8	110	5.2	171	2.2
56	8.1	111	96.9	181	100.0
60	5.1	112	15.3	182	7.4
61	19.6	113	25.5	183	98.5
62	5.2	114	4.2	184	7.1
63	11.4	119	3.3	185	28.1
72	3.5	121	15.8	187	3.9
73	55.4	122	6.8	216	12.1
74	23.0	123	12.1	217	58.3
75	44.2	124	4.7	218	17.1
76	6.3	125	2.4	219	83.0
77	42.7	131	2.3	220	13.0
78	9.9	133	6.4	221	28.4
83	30.5	134	2.1	222	4.6
84	3.9	135	3.6	223	8.3
85	42.7	143	6.6	252	9.3
86	3.4	145	26.6	254	16.1
87	19.8	146	22.6	256	9.6
88	2.1	147	33.2	258	3.4
89	4.5	148	16.1	288	2.9
95	3.8	149	11.1	290	5.0
96	17.0	150	3.1	292	4.1
97	5.2	156	14.3	294	2.3

Pesticide
CAS Name: Cyclohexane, 1,2,3,4,5,6-hexachloro-, γ-
Synonyms

Aalindan	Ben-Hex	Ent 7,796
Aficide	Benhexachlor	Entomoxan
Agrisol G-20	Bentox 10	Forlin
Agrocide	Benzene hexachloride	Gamma benzene hexachloride
Agrocide 2	Benzene hexachloride-γ-isomer	Gamacid
Agrocide 6G	γ-Benzene hexachloride	Gamaphex
Agrocide 7	Bexol	Gammahexa
Agrocide III	BHC	Gammahexane
Agrocide WP	Celanex	Gammalin
Agronexit	Chloresene	Gammalin 20
Aparasin	Codechine	Gammaterr
Aphtiria	DBH	HCCH
Aplidal	Detox 25	HCH
Arbitex	Devoran	γ-HCH
BBH	Dol granule	

continued overleaf

γ-BHC–*continued*

Synonyms–*continued*

- Heclotox
- Hexa
- Hexachloran
- γ-Hexachloran
- Hexachlorane
- γ-Hexachlorane
- γ-Hexachlorobenzene
- γ-Hexachlorocyclohexane
- 1,2,3,4,5,6-Hexachlorocyclohexane-γ-isomer
- γ-1,2,3,4,5,6-Hexachlorocyclohexane
- Hexatol
- Hexaverm
- Hexicide
- Hexyclan
- HGI
- Hortex
- Isotox
- Jacutin
- Kokotine
- Kwell
- Lendine
- Lentox
- Lidenal
- Lindafor
- Lindagam
- Lindane
- γ-Lindane
- Lindatox
- Lindosep
- Lintox
- Lorexane
- Malaoxon
- Milbol 49
- Mszycol
- NCI-C00204
- Neo-Scabicidol
- Nexen FB
- Nexit
- Nexit-Stark
- Nexol-E
- Nicochloran
- Omnitox
- Ovadziak
- Owadziak
- Quellada
- Sang gamma
- Silvanol
- Streunex
- TAP 85
- Tri-6
- Viton

CAS Registry No.: 58-89-9 (formerly 8007-42-9, 8073-23-2, 25897-48-7, 53529-37-6)
ROTECS Ref.: GV49000
Merck Index Ref.: 5341

Major Ions: 181, 183, 111, 219, 109, 217, 51, 73
EPA Ions: 183, 109, 181

Selected Bibliography

Ion kinetic energy spectra of some chlorinated insecticides, Safe, S., Hutzinger, O., Jamieson, W. D., Cook, M., *Org. Mass Spectrom.*, 7(2), 217–224 (1973).

Application of coupled gas chromatography–mass spectrometry in methods for the study and determination of pesticide residues and organic micropollutants in environmental and food materials, Mestres, R., Chevallier, C., Espinoza, C., Cornet, R., *Ann. Falsif. Exper. Chim.*, 70(751), 177–188 (1977).

δ-BHC $C_6H_6Cl_6$ (288)

Spectral Data

Mass	Abundance	Mass	Abundance	Mass	Abundance
47	4.1	51	55.2	60	5.5
48	3.8	52	5.5	61	23.5
49	17.8	55	8.5	62	6.6
50	33.9	56	4.1	63	14.2

Spectral Data–*continued*

Mass	Abundance	Mass	Abundance	Mass	Abundance
73	33.6	113	18.6	159	3.3
74	16.9	114	2.5	160	2.5
75	37.7	119	2.2	161	3.0
76	3.8	120	2.2	169	2.7
77	21.3	121	19.9	181	100.0
78	2.7	122	7.7	182	8.7
83	33.6	123	10.1	183	94.0
84	5.2	124	4.4	184	8.5
85	49.7	131	2.2	185	30.3
86	5.7	133	6.6	186	2.2
87	27.6	134	2.5	187	3.3
88	3.0	135	7.7	216	21.6
89	4.4	136	2.7	217	64.5
95	2.2	137	3.3	218	32.2
96	16.9	143	6.8	219	89.6
97	4.4	145	26.5	220	18.3
98	11.2	146	15.8	221	40.7
99	8.2	147	20.2	222	6.8
100	2.5	148	13.7	223	8.5
101	3.6	149	4.6	252	6.3
107	3.3	150	3.8	253	3.3
109	85.0	156	12.6	254	17.8
110	5.2	157	6.6	256	10.7
111	74.9	158	13.7	258	4.9
112	11.2				

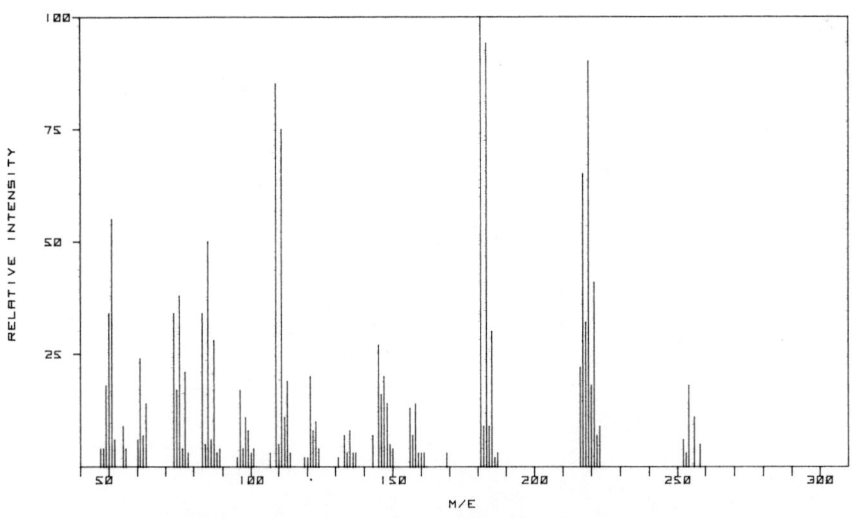

δ-BHC–*continued*

Pesticide
CAS Name: Cyclohexane, 1,2,3,4,5,6,-hexachloro-, δ-
Synonyms

δ-Benzenehexachloride	δ-Hexachlorocyclohexane
Ent 9,234	δ-1,2,3,4,5,6-Hexachlorocyclohexane
δ-HCH	δ-Lindane

CAS Registry No.: 319-86-8
ROTECS Ref.: GV45500

Major Ions: 181, 183, 219, 109, 111, 217, 51, 85
EPA Ions: 183, 109, 181

Selected Bibliography

Ion kinetic energy spectra of some chlorinated insecticides, Safe, S., Hutzinger, O., Jamieson, W. D., Cook, M., *Org. Mass Spectrom.*, 7(2), 217–224 (1973).

HEXACHLOROCYCLOPENTADIENE

This compound is an intermediate in the production of chlordane and endosulfan and is also used as a fumigant.

HEXACHLOROCYCLOPENTADIENE **C_5Cl_6 (270)**

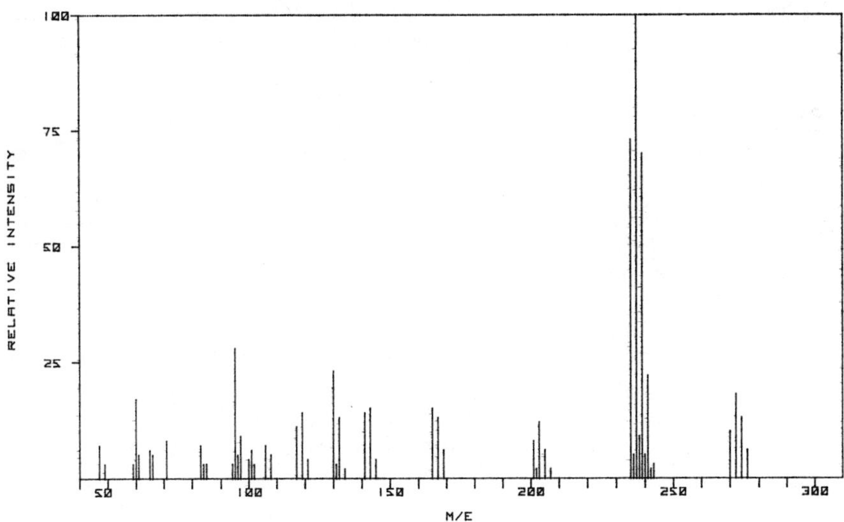

Spectral Data

Mass	Abundance	Mass	Abundance	Mass	Abundance
47	7.3	85	3.1	117	10.5
49	3.3	94	2.9	119	13.8
59	3.3	95	28.4	121	4.4
60	16.9	96	5.1	130	22.6
61	4.6	97	9.0	131	3.1
65	5.9	100	4.4	132	12.7
66	5.3	101	6.2	134	2.4
71	7.5	102	2.9	141	13.8
83	6.8	106	7.0	143	14.5
84	2.6	108	4.6	145	4.2

HEXACHLOROCYCLOPENTADIENE–*continued*

Spectral Data–*continued*

Mass	Abundance	Mass	Abundance	Mass	Abundance
165	14.5	207	2.4	241	22.2
167	13.4	235	73.0	242	2.2
169	5.9	236	5.1	243	3.3
201	8.4	237	100.0	270	9.7
202	2.2	238	9.2	272	17.8
203	12.1	239	69.7	274	12.5
205	6.2	240	5.3	276	5.9

Base–neutral extractable
CAS Name: 1,3-Cyclopentadiene, 1,2,3,4,5,5-hexachloro-
Synonyms

- C-56
- Graphlox
- 1,2,3,4,5,5-Hexachloro-1,3-cyclopentadiene
- HRS 1655
- PCL
- Perchlorocyclopentadiene

CAS Registry No.: 77-47-4
ROTECS Ref.: GY12250

Major Ions: 237, 235, 239, 95, 130, 241, 272, 60
EPA Ions: 237, 235, 272

Selected Bibliography

High-resolution mass spectrometry matrix analysis of environmental samples, Stauffer, J. L., Levins, P. L., Oberholtzer, J. E., *Carcinogens–A Comprehensive Survey*, vol 3: *Polynuclear Aromatic Hydrocarbons*, Raven Press, New York (1978), pp. 89–95.

ISOPHORONE

This solvent is manufactured by the condensation of acetone in the presence of sodium hydroxide.

ISOPHORONE $C_9H_{14}O$ (138)

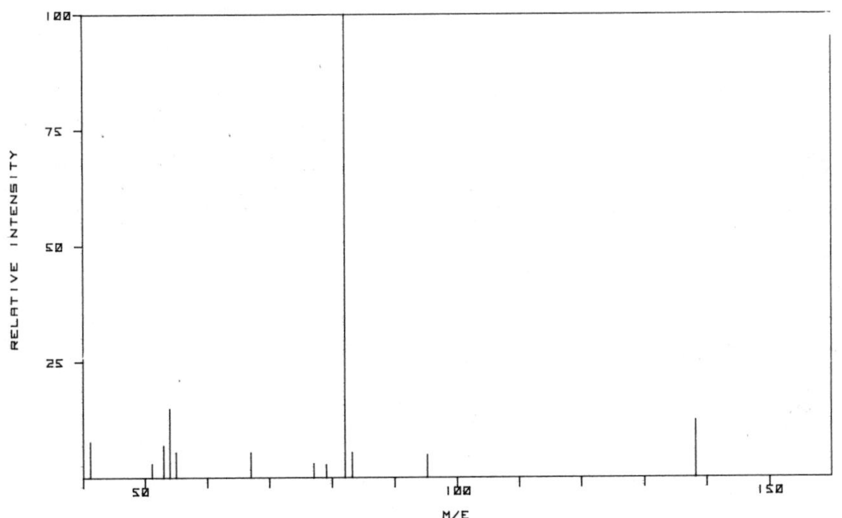

Spectral Data

Mass	Abundance	Mass	Abundance	Mass	Abundance
41	7.7	55	5.3	82	100.0
51	2.9	67	5.3	83	5.3
53	6.9	77	2.9	95	4.8
54	14.8	79	2.7	138	12.2

Base–neutral extractable
CAS Name: 2-Cyclohexen-1-one, 3,5,5-trimethyl-

ISOPHORONE–*continued*

Synonyms

Isoacetophorone	α-Isophorone
Isoforon	1,1,3-Trimethyl-3-cyclohexene-5-one
Isophoron	3,5,5-Trimethyl-2-cyclohexene-1-one
α-Isophoron	3,5,5-Trimethyl-2-cyclohexenone

CAS Registry No.: 78-59-1
ROTECS Ref.: GW77000

Major Ions: 82, 54, 138, 41, 53, 83, 67, 55
EPA Ions: 82, 95, 138

NAPHTHALENE

Naphthalene is the most abundant constituent of coal tar. It is used as a moth repellant and insect fumigant. This compound is used in the manufacture of celluloid, dyes, hydronaphthalenes, lampblack, smokeless powder, and synthetic resins. It is employed as a topical and intestinal antiseptic and anthelmintic.

NAPHTHALENE **$C_{10}H_8$ (128)**

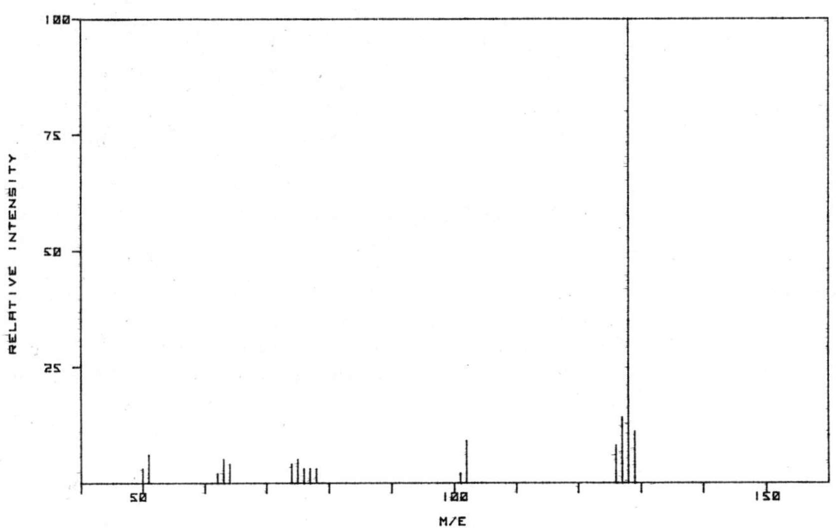

Spectral Data

Mass	Abundance	Mass	Abundance	Mass	Abundance
50	3.4	75	4.9	102	8.9
51	5.8	76	2.8	126	7.8
62	2.3	77	3.0	127	14.0
63	4.7	78	2.7	128	100.0
64	4.1	101	2.2	129	11.4
74	4.2				

NAPHTHALENE–*continued*

Base–neutral extractable
CAS Name: Naphthalene
Synonyms

Albocarbon	Naphthalin
Camphor tar	Naphthaline
Dezodorator	Naphthene
Moth balls	Tar camphor
Moth flakes	White tar

CAS Registry No.: 91-20-3
ROTECS Ref.: QJ05250
Merck Index Ref.: 6194

Major Ions: 128, 127, 129, 102, 126, 51, 75, 63
EPA Ions: 128, 129, 127

Selected Bibliography

Mass spectra of naphthalene and its methyl derivatives, Poponova, R. V., Lukashenko, I. M., Polyakova, A. A., Rang, S., Eisen, O., *Zh. Org. Khim.*, 7(10), 2032–2038 (1971).

Use of mass spectrometry for the analysis of oily substances used for wood preservation, Dolansky, V., Komora, F., *Sb. Vys. Sk. Chem.-Technol. Praze. Technol. Paliv.*, **D30**, 329–334 (1974).

A survey of the molecular nature of primary and secondary components of particles in urban air by high-resolution mass spectrometry, Cronn, D. R., Charlson, R. J., Knights, R. L., Crittenden, A. L., Appel, B. R., *Atmos. Environ.*, **11**(10), 929–937 (1977).

Gas chromatography–mass spectrometry of simulated arson residue using gasoline as an accelerant, Mach, M. H., *J. Forensic Sci.*, **22**(2), 348–357 (1977).

Selection of a method for determining the composition of aromatic hydrocarbons in high-boiling petroleum fractions, Siryuk, A. G., Barabadze, Sh. Sh., *Khim. Technol. Topl. Masel*, (10), 54–56 (1977).

NITROBENZENE

Nitrobenzene is used in the manufacture of aniline, pyroxylin compounds, shoe polishes, and soaps, and in the refining of lubricating oils. It was formerly used as an insecticide.

NITROBENZENE $C_6H_5NO_2$ (123)

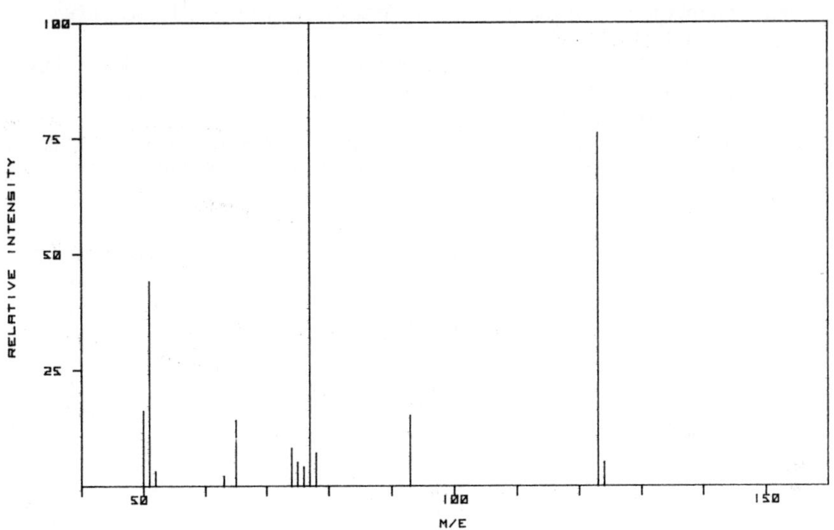

Spectral Data

Mass	Abundance	Mass	Abundance	Mass	Abundance
50	16.0	74	7.9	78	6.7
51	43.8	75	4.6	93	14.8
52	2.5	76	3.7	123	75.6
63	2.1	77	100.0	124	5.4
65	14.0				

Base–neutral extractable
CAS Name: Benzene, nitro-
Synonyms

- C.I. solvent black 5
- Essence of mirbane
- Essence of myrbane
- Mirbane oil
- Nigrosine spirit soluble B
- Nitrobenzol
- Nitrobenzol, liquid
- Oil of mirbane
- Oil of myrbane

NITROBENZENE–*continued*

CAS Registry No.: 98-95-3
ROTECS Ref.: DA64750
Merck Index Ref.: 6409

Major Ions: 77, 123, 51, 50, 93, 65, 74, 78
EPA Ions: 77, 123, 65

Selected Bibliography

Interpretation of mass spectra by simple LCAO-MO calculation. II. Derivatives of benzene and chlorobenzene, Tajima, S., Niwa, Y., Wasada, N., Tsuchiya, T., *Bull. Chem. Soc. Jpn.*, **45**(4), 1250–1251 (1972).

Field mass spectra of nitro compounds, Fileleeva, L. I., Aleksankin, M. M., Koval'chuk, V. N., Korostyshevskii, I. Z., Mischanchuk, B. G., Grom, V. V., *Teor. Eksp. Khim.*, **11**(3), 342–348 (1975).

Doubly charged ion mass spectra of monosubstituted aromatic compounds, Sakurai, H., Tatematsu, A., Nakata, H., *Bull. Chem. Soc. Jpn.*, **49**(10), 2800–2801 (1976).

NITROPHENOLS

4,6-Dinitro-*o*-cresol, 2,4-dinitrophenol, 2-nitrophenol, and 4-nitrophenol are the compounds included in this category.

4,6-Dinitro-*o*-cresol is a potent insecticide, but is little used for this purpose since it is phytotoxic. However, it is employed as an ovicidal spray on dormant fruit trees, as a selective herbicide, and for thinning fruit on apple trees. It is explosive when dry. The sodium and potassium salts are water soluble.

2,4-Dinitrophenol is used as a wood preservative and in the manufacture of dyes.

2- and 4-nitrophenols are used for the manufacture of a wide variety of industrial chemicals. The latter compound is formed by the hydrolysis of the insecticide parathion.

4,6-DINITRO-*o*-CRESOL $C_7H_6N_2O_5$ (198)

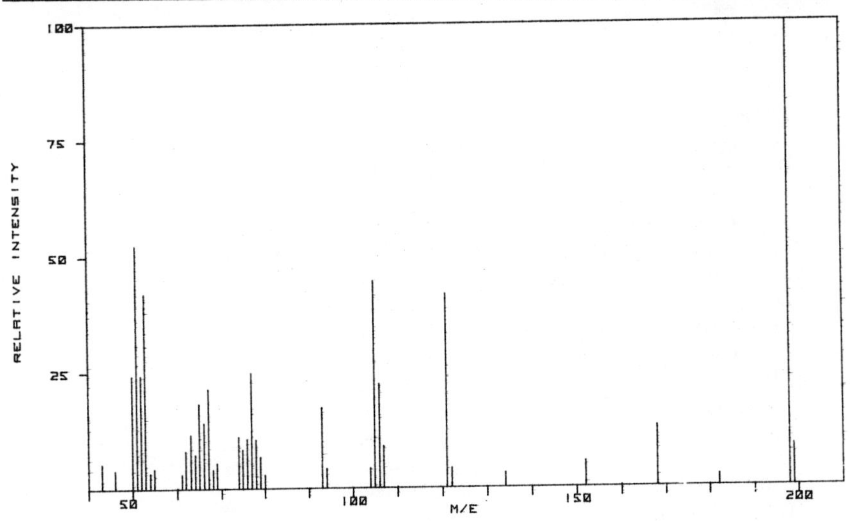

Spectral Data

Mass	Abundance	Mass	Abundance	Mass	Abundance
43	5.2	53	41.9	63	11.4
46	3.8	54	3.2	64	7.1
50	24.2	55	4.1	65	18.1
51	52.3	61	2.9	66	13.9
52	24.2	62	7.9	67	21.3

4,6-DINITRO-*o*-CRESOL–*continued*

Spectral Data–*continued*

Mass	Abundance	Mass	Abundance	Mass	Abundance
68	3.9	80	2.7	122	3.9
69	5.3	93	17.2	134	2.9
74	10.9	94	4.0	152	5.3
75	8.2	104	4.0	168	12.8
76	10.4	105	44.5	182	2.2
77	24.7	106	22.2	198	100.0
78	10.2	107	8.7	199	8.5
79	6.5	121	41.5		

Acid extractable
CAS Name: *o*-Cresol, 4,6-dinitro-
Synonyms

Antinonin
Antinonnin
Arborol
Capsine
Degrassan
Dekrysil
Detal
Dillex
Dinitro
Dinitrocresol
Dinitro-*o*-cresol
2,4-Dinitro-*o*-cresol
3,5-Dinitro-*o*-cresol
3,5-Dinitro-2-hydrotoluene
3,5-Dinitro-2-hydroxytoluene
Dintrol
Dinitromethylcyclohexyltrienol
2,4-Dinitro-6-methylphenol
Dinoc
Dinurania
Ditrosol
DN
DNC
DN-Dry Mix No. 2
DNOC
Effusan
Effusan 3436
Elgetol
Elgetol 30
Elipol
Extrar
Hedolit
Hedolite
K III
K IV
Krenite
Krezotol 50
Lipan
2-Methyl-4,6-dinitrophenol
6-Methyl-2,4-dinitrophenol
Nitrador
Nitrofan
Prokarbol
Rafex
Rafex 35
Raphatox
Sandolin
Sandolin A
Selinon
Sinox
Trifocide
Winterwash

CAS Registry No.: 534-52-1 (formerly 8068-73-3, 8071-51-0, 37359-43-6, 53240-95-2)
ROTECS Ref.: GO96250
Merck Index Ref.: 3275

Major Ions: 198, 51, 105, 53, 121, 77, 52, 50
EPA Ions: 198, 182, 77

2,4-DINITROPHENOL $C_6H_4N_2O_5$ (184)

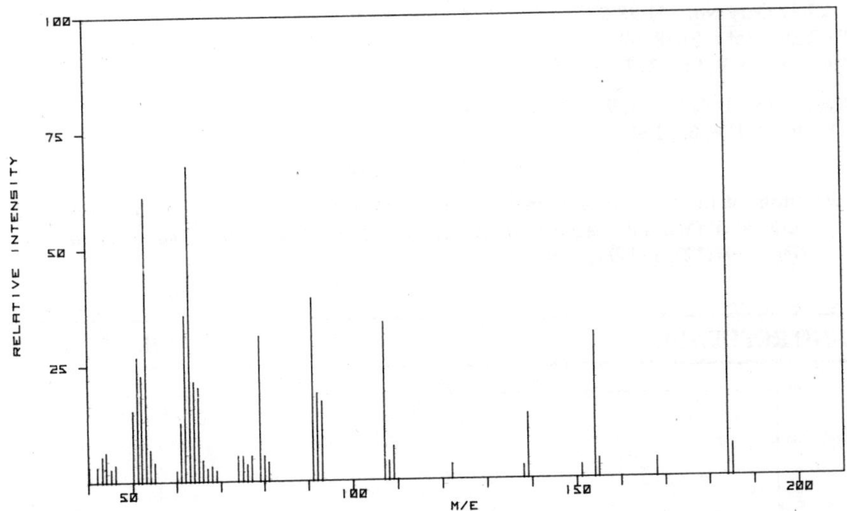

Spectral Data

Mass	Abundance	Mass	Abundance	Mass	Abundance
40	3.1	63	67.9	92	18.8
42	3.1	64	21.4	93	17.0
43	5.4	65	20.1	107	33.9
44	6.3	66	4.5	108	4.0
45	2.7	67	2.7	109	7.1
46	3.6	68	3.1	122	3.1
50	15.2	69	2.2	138	2.7
51	26.8	74	5.4	139	13.8
52	22.8	75	5.4	151	2.7
53	61.2	76	3.6	154	31.3
54	6.7	77	5.4	155	4.0
55	4.0	79	31.3	168	4.0
60	2.2	80	5.4	184	100.0
61	12.5	81	4.0	185	6.7
62	35.7	91	39.3		

Acid extractable
CAS Name: Phenol, 2,4-dinitro-
Synonyms

- Aldifen
- Chemox PE
- α-Dinitrophenol
- 2,4-DNP
- Fenoxyl carbon N
- 1-Hydroxy-2,4-dinitrobenzene
- Nitro Kleenup
- Nitrophen
- Nitrophene
- NSC 1,532
- Solfo black B
- Solfo black BB
- Solfo black 2B supra
- Solfo black G
- Solfo black SB
- Tertrosulphur black PB
- Tertrosulphur PBR

2,4-DINITROPHENOL—*continued*

CAS Registry No.: 51-28-5
ROTECS Ref.: SL28000
Merck Index Ref.: 3277

Major Ions: 184, 63, 53, 91, 62, 107, 154, 79
EPA Ions: 184, 63, 154

Selected Bibliography

Mechanism of the reaction of aromatic compounds with formaldehyde in concentrated sulfuric acid (Marquis reagent), Rehse, K., Kawerau, H. G., *Arch. Pharm.* (*Weinheim, Ger.*), **307**(12), 934–942 (1974).

2-NITROPHENOL $C_6H_5NO_3$ (139)

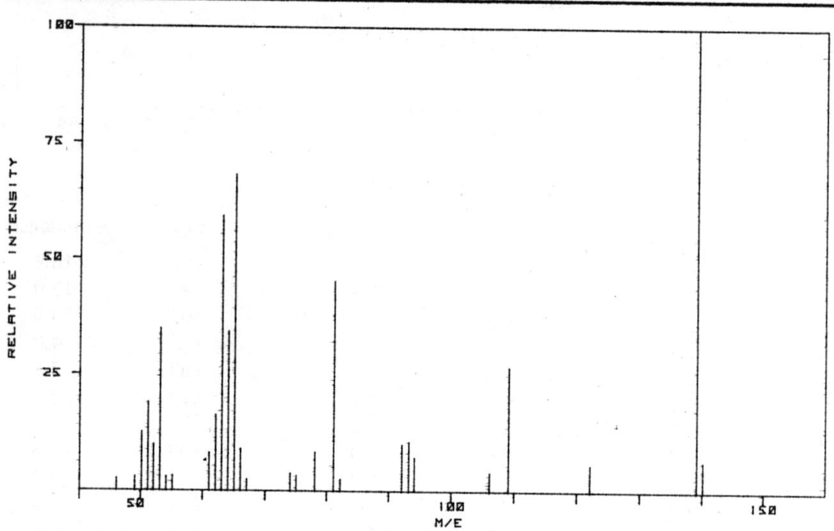

Spectral Data

Mass	Abundance	Mass	Abundance	Mass	Abundance
40	3.5	62	16.2	82	2.5
46	2.5	63	59.2	92	9.9
49	2.9	64	34.2	93	10.6
50	12.5	65	68.0	94	7.2
51	19.0	66	8.9	106	4.0
52	9.8	67	2.4	109	26.7
53	34.8	74	3.7	122	5.6
54	2.9	75	3.2	139	100.0
55	3.1	78	8.2	140	6.4
61	8.1	81	45.2		

Acid extractable
CAS Name: Phenol, *o*-nitro-
Synonyms
2-Hydroxynitrobenzene
o-Hydroxynitrobenzene
o-Nitrophenol

CAS Registry No.: 88-75-5
ROTECS Ref.: SM21000
Merck Index Ref.: 6442

Major Ions: 139, 65, 63, 81, 53, 64, 109, 51
EPA Ions: 139, 65, 109

Selected Bibliography

Temperature dependence of mass spectra and conformational transformation of ortho-substituted phenols in the gas phase, Bogolyubov, G. M., Gal'perin, Ya. V., Petrov, A. A., *Zh. Obshch. Khim.*, **46**(2), 336–340 (1976).

4-NITROPHENOL — $C_6H_5NO_3$ (139)

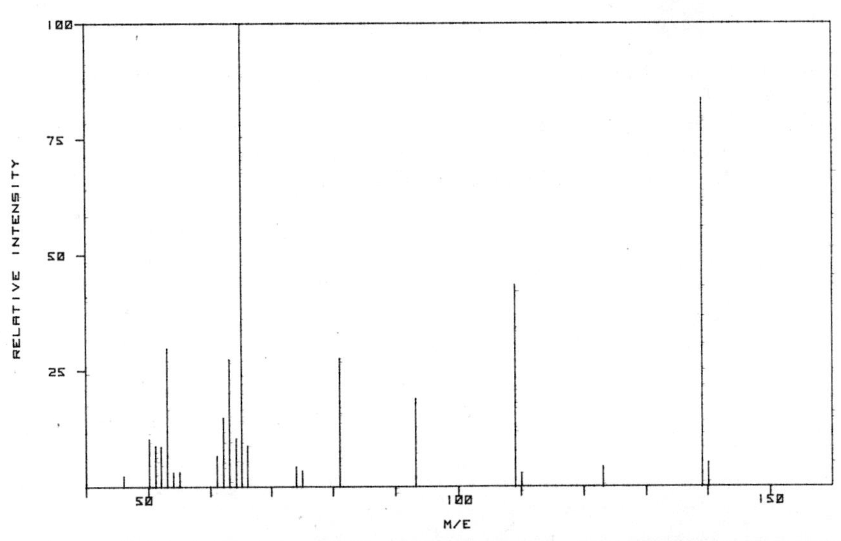

4-NITROPHENOL—*continued*

Spectral Data

Mass	Abundance	Mass	Abundance	Mass	Abundance
40	2.8	61	6.5	81	27.6
46	2.3	62	14.8	93	18.9
50	10.2	63	27.3	109	43.4
51	8.7	64	10.3	110	2.9
52	8.5	65	100.0	123	4.2
53	29.7	66	8.7	139	83.6
54	3.0	74	4.2	140	5.1
55	3.1	75	3.3		

Acid extractable
CAS Name: Phenol, *p*-nitro-
Synonyms
- 4-Hydroxynitrobenzene
- *p*-Hydroxynitrobenzene
- Niphen
- *p*-Nitrophenol
- PNP

CAS Registry No.: 100-02-7
ROTECS Ref.: SM22750
Merck Index Ref.: 6443

Major Ions: 65, 139, 109, 53, 81, 63, 93, 62
EPA Ions: 65, 139, 109

Selected Bibliography

Mass identified mobility spectra of *p*-nitrophenol and reactant ions in plasma chromatrograpy, Karasek, F. W., Kim, S. H., Hill, H. H., Jr., *Anal. Chem.*, 48(8), 1133–1137 (1976).

Protonation, ethylation, and allylation of substituted nitrobenzenes in the gas phase. A study by ion kinetic energy spectrometry and chemical ionization, Kruger, T. L., Flammang, R., Litton, J. F., Cooks, R. G., *Tetrahedron Lett.*, (50), 4555–4558 (1976).

NITROSAMINES

N-Nitrosodimethylamine, *N*-nitrosodiphenylamine, and *N*-nitrosodi-n-propylamine are the compounds included in this category.

N-Nitrosodiphenylamine is used as accelerator in vulcanizing rubber.

During the 1960's there were many reports of the presence of nitrosamines in foodstuffs, with identification based on GC retention time and polarography. GC-MS later demonstrated that the compounds suspected of being nitrosamines were actually alkyl triazines. More recently, however, SIM during GC-MS has shown that nitrosamines are present in small amounts in some foodstuffs, including cured meats and smoked and salted fish. Unequivocal evidence for the presence of nitrosamines in tobacco smoke and urban air from several locations has also been obtained.

N-Nitrosodimethylamine tails badly during gas chromatography under standard E.P.A. conditions. *N*-Nitrosodiphenylamine decomposes to afford diphenylamine during gas chromatography. The mass spectrum of the latter compound is reported here.

N-NITROSODIMETHYLAMINE $C_2H_6N_2O$ (74)

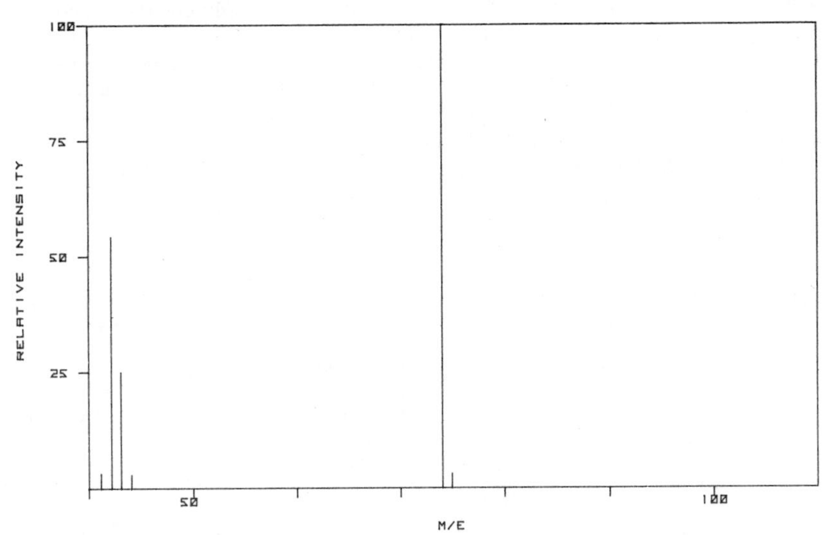

N-NITROSODIMETHYLAMINE–*continued*

Spectral Data

Mass	Abundance	Mass	Abundance	Mass	Abundance
40	3.8	43	24.9	74	100.0
41	3.1	44	2.8	75	3.0
42	54.1				

Base–neutral extractable
CAS Name: Dimethylamine, *N*-nitroso-
Synonyms

N,*N*-Dimethylnitrosamine
Dimethylnitrosamine
DMN
DMNA
N-Methyl-*N*-nitrosomethanamine
NDMA

CAS Registry No.: 62-75-9
ROTECS Ref.: IQ05250
Merck Index Ref.: 6458

Major Ions: 74, 42, 43, 40, 41, 75, 44
EPA Ions: 42, 74, 44

Selected Bibliography

Mass spectrometry of volatile derivatives. V. *N*-Dialkylnitrosamines, Saxby, M. J., *J. Assoc. Offic. Anal. Chem.*, **55**(1), 9–12 (1972).

Use of isomers in the detection and estimation of volatile nitrosamines by combined high-resolution mass spectrometry–gas chromatography, Crathorne, B., Edwards, M. W., Jones, N. R., Walters, C. L., Woolford, G., *J. Chromatogr.*, **115**(1), 213–217 (1975).

Pyridine catalyzed reaction of volatile *N*-nitrosamines with heptafluorobutyric anhydride, Gough, T. A., Sugden, K., Webb, K. S., *Anal. Chem.*, **47**(3), 509–512 (1975).

Application of new mass spectrometry techniques for gas chromatography/mass spectrometry routine analysis in environmental chemistry, food control and related fields, Naegeli, P., Egli, H. P., *Adv. Mass Spectrom.*, **7B**, 1713–1720 (1978).

Mass-spectrometric analysis of *N*-nitrosamines, Pokrovskii, A. A., Kostyukovskii, Ya. L., Melamed, D. B., Medvedev, F. A., *Zh. Anal. Khim.*, **33**(5), 970–974 (1978).

N-NITROSODIPHENYLAMINE $C_{12}H_{10}N_2O$ (198)

Spectral Data for Diphenylamine (see p. 139)

Mass	Abundance	Mass	Abundance	Mass	Abundance
50	4.6	77	100.0	153	2.7
51	23.7	78	7.3	182	21.9
63	2.1	105	18.4	183	2.9
64	2.2	152	5.3		

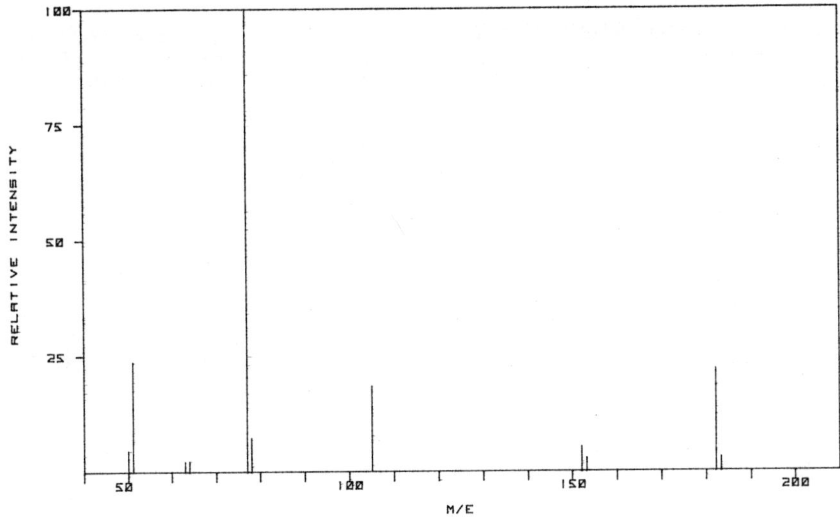

Base–neutral extractable
CAS Name: Diphenylamine, *N*-nitroso-
Synonyms

Diphenylnitrosamine	Retarder J
NCI-C02880	Vulcatard
N-Nitroso-*N*-phenylaniline	Vulcatard A
4-Nitroso-*N*-phenylbenzenamine	Vulkalent A
Redax	Vultrol

CAS Registry No.: 86-30-6
ROTECS Ref.: JJ98000
Merck Index Ref.: 6460

Major Ions: 77, 51, 182, 105, 78, 152, 50, 183
EPA Ions: 169, 168, 167
} for diphenylamine (see p. 139).

Selected Bibliography

Synthesis and spectral characteristics of some nitrosoanilines and their possible mode of action as curing agents for various types of rubbers, Potts, K. T., Kane, J., McKeough, D., D'Amico, J. J., *Rubber Chem. Technol.*, 47(2), 289–302 (1974).

Gas phase protonolysis reactions. Chemical ionization mass spectrometry of *N*-nitrosamines, Fish, R. H., Holmstead, R. L., Gaffield, W., *Tetrahedron*, **32**(22), 2689–2692 (1976).

Chemical ionization–mass spectrometry of nitrosamines, Gaffield, W., Fish, R. H., Holmstead, R. L., Poppiti, J., Yergey, A. L., *IARC Scientific Publications*, vol. 14: *Environ. N-Nitroso Compd. Anal. Form., Proc. Work. Conf., 1975* (1976), pp. 11–20.

N-NITROSODI-n-PROPYLAMINE $C_6H_{14}N_2O$ (130)

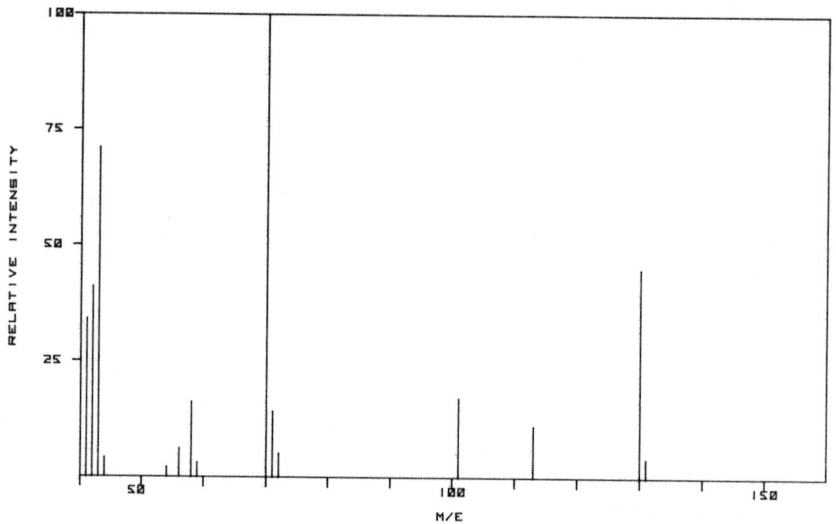

Spectral Data

Mass	Abundance	Mass	Abundance	Mass	Abundance
41	33.9	56	5.8	72	5.0
42	41.4	58	15.8	101	16.6
43	70.7	59	3.2	113	10.8
44	3.9	70	100.0	130	44.8
54	2.1	71	14.3	131	3.5

Base–neutral extractable
CAS Name: Dipropylamine, N-nitroso-
Synonyms

- Dipropylnitrosamine
- Di-n-propylnitrosamine
- N,N-Dipropylnitrosamine
- DPN
- Nitrosodipropylamine
- N-Nitroso-N-dipropylamine
- N-Nitroso-N-propyl-1-propanamine

CAS Registry No.: 621-64-7
ROTECS Ref.: JL97000

Major Ions: 70, 43, 130, 42, 41, 101, 58, 113
EPA Ions: 130, 42, 101

Selected Bibliography

Mass spectrometry of volatile derivatives. V. *N*-Dialkylnitrosamines, Saxby, M. J., *J. Assoc. Offic. Anal. Chem.*, **55**(1), 9–12 (1972).

Use of isomers in the detection and estimation of volatile nitrosamines by combined high-resolution mass spectrometry–gas chromatography, Crathorne, B., Edwards, M. W., Jones, N. R., Walters, C. L., Woolford, G., *J. Chromatogr.*, **115**(1), 213–217 (1975).

Gas phase protonolysis reactions. Chemical ionization mass spectrometry of *N*-nitrosamines, Fish, R. H., Holmstead, R. L., Gaffield, W., *Tetrahedron*, **32**(22), 2689–2692 (1976).

Chemical ionization–mass spectrometry of nitrosamines, Gaffield, W., Fish, R. H., Holmstead, R. L., Poppiti, J., Yergey, A. L., *IARC Scientific Publications*, vol. 14 *Environ. N-Nitroso Compd. Anal. Form., Proc. Work. Conf., 1975* (1976), pp. 11–20.

Application of new mass spectrometry techniques for gas chromatography/mass spectrometry routine analysis in environmental chemistry, food control and related fields, Naegeli, P., Egli, H. P., *Adv. Mass Spectrom.*, **7B**, 1713–1720 (1978).

PENTACHLOROPHENOL

This compound is used for the protection of timber against fungi, termites, and other wood borers. It has also been used as a herbicide and preharvest defoliant. The sodium and zinc salts are water soluble.

PENTACHLOROPHENOL C_6HCl_5O (264)

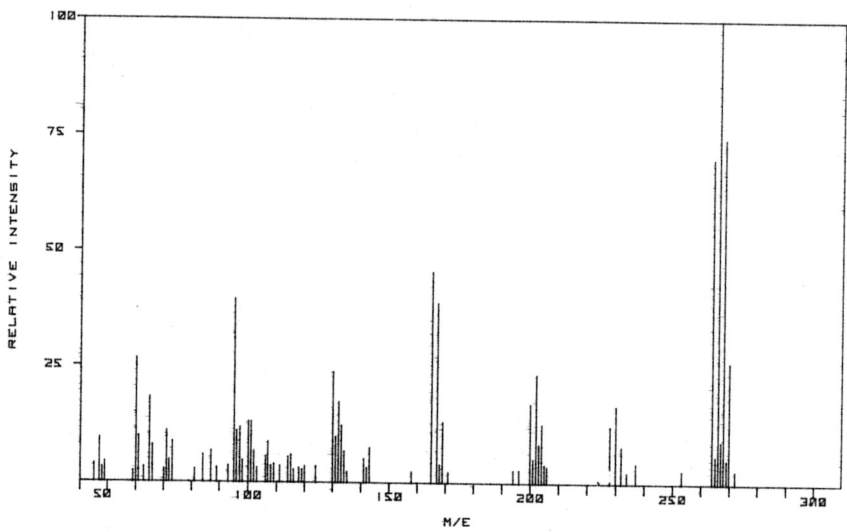

Spectral Data

Mass	Abundance	Mass	Abundance	Mass	Abundance
45	4.0	65	18.2	87	6.7
47	9.5	66	7.9	89	3.2
48	3.2	70	2.8	93	3.6
49	4.3	71	11.1	95	39.5
59	2.4	72	4.7	96	11.1
60	26.5	73	8.7	97	11.9
61	9.9	81	2.8	98	4.7
63	3.2	84	5.9	100	13.0

Spectral Data—*continued*

Mass	Abundance	Mass	Abundance	Mass	Abundance
101	13.0	133	12.3	204	12.6
102	6.7	134	6.7	205	4.0
103	3.2	135	2.4	206	3.6
106	5.5	141	5.1	228	12.3
107	8.7	142	3.2	230	16.6
108	3.6	143	7.5	232	7.9
109	4.0	158	2.4	234	2.4
111	3.6	165	45.5	237	4.3
114	5.5	167	38.7	253	2.8
115	5.9	168	4.0	264	70.0
116	2.8	169	13.0	265	5.9
118	3.2	171	2.4	266	100.0
119	2.8	194	2.8	267	9.1
120	3.6	196	2.8	268	74.3
124	3.6	200	17.0	269	5.1
130	23.7	201	5.1	270	26.1
131	9.9	202	23.3	272	2.8
132	17.4	203	8.3		

Acid extractable
CAS Name: Phenol, pentachloro-
Synonyms

Chem-tol	NCI-C54933	Permacide
Chlorophen	PCP	Permagard
Dowicide G	Penchlorol	Permasan
Dowicide 7	Penta	Permatox
Durotox	Pentachlorophenate	Permite
EP 30	2,3,4,5,6-Pentachlorophenol	Santobrite
Fungifen	Pentacon	Santophen
Glazd	Penta-kil	Santophen 20
Grundier Arbezol	Pentanol	Sinituho
Lauxtol	Pentasol	Term-i-trol
Lauxtol A	Penwar	Thompson's wood fix
Liroprem	Peratox	Weedone

CAS Registry No.: 87-86-5
ROTECS Ref.: SM63000
Merck Index Ref.: 6901

Major Ions: 266, 268, 264, 165, 95, 167, 60, 270
EPA Ions: 266, 264, 268

PENTACHLOROPHENOL–*continued*

Selected Bibliography

Gas–liquid chromatography as an analytical tool in the production of pentachlorophenol and hexachlorophenol, Svec. P., Zbirovsky, M., *Sb. Vys. Sk. Chem.-Technol. Praze, Org. Chem. Technol.*, **C21**, 39–43 (1974).

Anomalous behavior of polychloro alicyclic ketones in gas chromatography and gas chromatography–mass spectrometry, Svec, P., Kubelka, V., *Fresenius' Z. Anal. Chem.*, **277**(2), 113–118 (1975).

Screening by negative chemical ionization mass spectrometry for environmental contamination with toxic residues: application to human urines, Dougherty, R. C., Piotrowska, K., *Proc. Natl. Acad. Sci. USA*, **73**(6), 1777–1781 (1976).

GC/MS analysis of organic compounds in domestic wastewaters, Garrison, A. W., Pope, J. D., Allen, F. R., *Identif. Anal. Org. Pollut. Water, (Chem. Congr. North Am. Cont.), 1st,* Ann Arbor Sci., Ann Arbor, Michigan (1976), pp. 517–556.

PESTICIDES AND METABOLITES

Aldrin, dieldrin, and chlordane are the compounds included in this category. Many additional pesticides and metabolites are included elsewhere.

Aldrin and diedrin are both potent contact insecticides. Dieldrin (an isomer of endrin, *q.v.*) are formed from aldrin by autoxidation or metabolism.

Chlordane is a mixture containing α-chlordane, γ-chlordane, chlordene, heptachlor, hexachlorocyclopentadiene, and other compounds. Its use as an insecticide is severely restricted in the U.S.

ALDRIN $C_{12}H_8Cl_6$ (362)

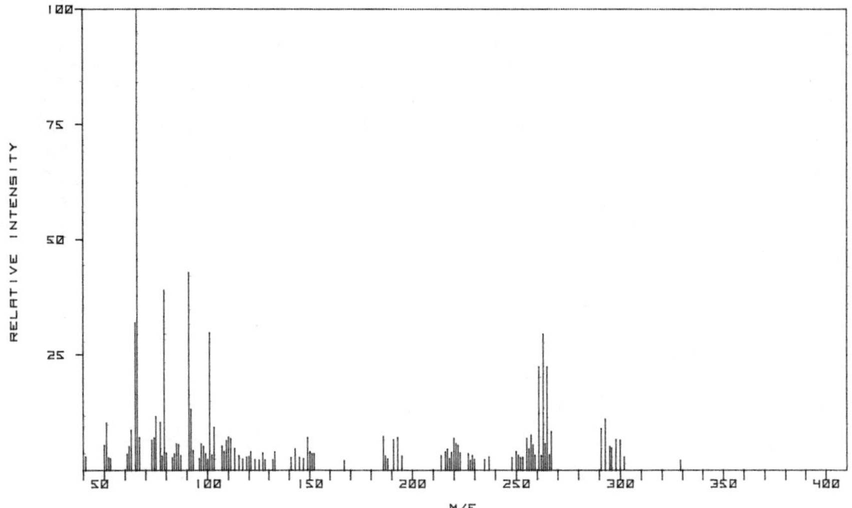

Spectral Data

Mass	Abundance	Mass	Abundance	Mass	Abundance
41	2.8	67	6.9	85	5.6
50	5.3	73	6.4	86	5.4
51	10.1	74	6.8	87	3.0
52	2.6	75	11.5	91	42.7
53	2.4	77	10.2	92	13.1
61	3.4	78	2.9	93	4.2
62	5.0	79	38.9	96	2.4
63	8.5	80	3.6	97	5.6
65	1.9	83	2.6	98	5.0
66	100.0	84	3.4	99	3.4

ALDRIN–*continued*

Spectral Data–*continued*

Mass	Abundance	Mass	Abundance	Mass	Abundance
100	2.2	150	3.9	248	2.6
101	29.6	151	3.4	250	3.9
102	3.1	152	3.5	251	3.0
103	9.1	167	2.0	252	2.6
107	5.1	186	7.2	253	2.7
108	3.9	187	2.9	255	6.7
109	6.3	188	2.3	256	4.4
110	7.0	191	6.5	257	7.4
111	6.7	193	6.9	258	5.3
113	4.6	195	3.0	259	3.1
115	3.0	214	3.0	261	22.3
117	2.3	216	3.8	262	3.0
119	2.7	217	4.4	263	29.4
120	2.8	218	2.4	264	5.6
121	3.9	219	3.8	265	22.3
123	2.2	220	6.8	266	3.2
125	2.1	221	5.7	267	8.2
127	3.6	222	5.3	291	8.9
128	2.1	223	3.6	293	10.9
132	2.2	227	3.5	295	5.0
133	3.8	228	2.0	296	4.7
141	2.6	229	3.0	298	6.5
143	4.5	230	2.3	300	6.4
145	2.7	235	2.2	302	2.7
147	2.4	237	2.8	329	2.0
149	6.9				

Pesticide

CAS Name: 1,4:5,8-Dimethanonaphthalene, 1,2,3,4,10,10-hexachloro-1,4,4a,5,8,8a-hexahydro-, *endo,exo-*

Synonyms

- Aldrex
- Aldrite
- Aldrosol
- Compound 118
- Drinox
- Ent 15,949
- 1,2,3,4,10,10-Hexachloro-1,4,4a,5,8,8a-hexahydro-1,4:5,8-dimethanonaphthalene
- Hexachlorohexahydro-*endo-exo*-dimethanonaphthalene
- 1,2,3,4,10,10-Hexachloro-1,4,4a,5,8,8a-hexahydro-1,4-*endo-exo*-5,8-dimethanonaphthalene
- 1,2,3,4,10,10-Hexachloro-1,4,4a,5,8,8a-hexahydro-*exo*-1,4-*endo*-5,8-dimethanonaphthalene
- HHDN
- NCI-C00044

Synonyms—*continued*

Octalene
SD 2794
Seedrin

CAS Registry No.: 309-00-2 (formerly 3714-23-6, 6851-31-6, 24562-14-9, 34487-55-3)
ROTECS Ref.: IO21000
Merck Index Ref.: 220

Major Ions: 66, 91, 79, 65, 101, 263, 265, 261
EPA Ions: 66, 263, 220

Selected Bibliography

Positive chemical ionization mass spectra of polycyclic chlorinated pesticides, Biros, F. J., Dougherty, R. C., Dalton, J., *Org. Mass Spectrom.*, **6**(11), 1161–1169 (1972).

Negative chemical ionization mass spectra of polycyclic chlorinated insecticides, Dougherty, R. C., Dalton, J., Biros, F. J., *Org. Mass Spectrom.*, **6**(11), 1171–1181 (1972).

Fate of carbon-14-labeled aldrin in potatoes and soil under outdoor conditions, Klein, W., Kohli, J., Weisgerber, I., Korte, F., *J. Agr. Food Chem.*, **21**(2), 152–156 (1973).

Negative chemical ionization mass spectrometry. Chloride attachment spectra, Tannenbaum, H. P., Roberts, J. D., Dougherty, R. C., *Anal. Chem.*, **47**(1), 49–54 (1975).

Application of coupled gas chromatography–mass spectrometry in methods for the study and determination of pesticide residues and organic micropollutants in environmental and food materials, Mestres, R., Chevallier, C., Espinoza C., Cornet, R., *Ann. Falsif. Exper. Chim.*, **70**(751), 177–188 (1977).

DIELDRIN $C_{12}H_8Cl_6O$ (378)

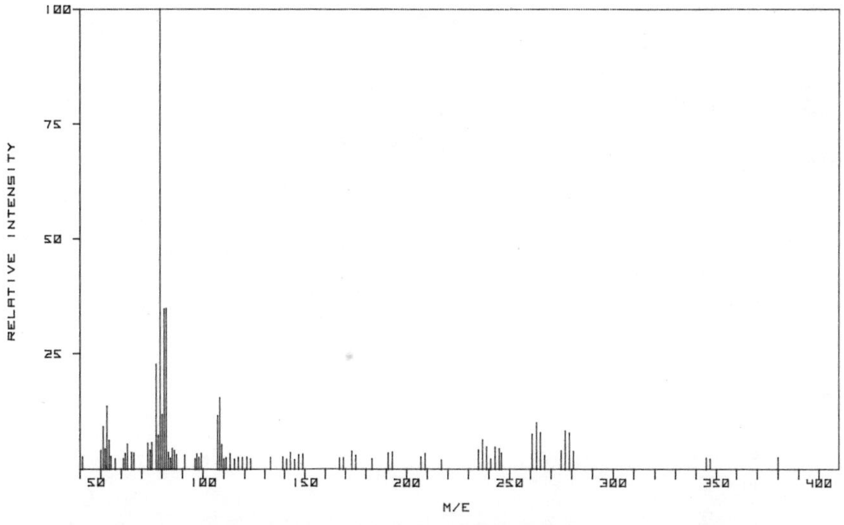

DIELDRIN–*continued*

Spectral Data

Mass	Abundance	Mass	Abundance	Mass	Abundance
41	2.6	87	3.0	173	3.9
50	4.0	91	2.9	175	3.0
51	9.1	96	2.1	183	2.3
52	4.3	97	3.2	191	3.6
53	13.6	98	2.5	193	3.7
54	6.1	99	3.3	207	2.7
55	2.7	107	11.5	209	3.4
57	2.2	108	15.4	217	2.0
61	2.2	109	5.2	235	4.2
62	3.3	110	2.1	237	6.4
63	5.3	111	2.4	239	4.8
65	3.6	113	3.3	241	2.2
66	3.4	115	2.1	243	4.8
73	5.5	117	2.4	245	4.4
74	4.0	119	2.4	246	3.5
75	5.7	121	2.5	261	7.6
77	22.6	123	2.2	263	10.1
78	7.3	133	2.5	265	7.9
79	100.0	139	2.6	267	2.9
80	11.7	141	2.2	275	4.0
81	34.7	143	3.6	277	8.3
82	34.8	145	2.0	279	7.8
83	3.6	147	3.0	281	3.9
84	2.3	149	3.2	345	2.4
85	4.4	167	2.4	347	2.2
86	3.9	169	2.4	380	2.4

Pesticide

CAS Name: 1,4:5,8-Dimethanonaphthalene, 1,2,3,4,10,10-hexachloro-6,7-epoxy-1,4,4a,5,6,7,8,8a-octahydro-, *endo,exo-*

Synonyms

- Aldrin epoxide
- Alvit
- Alvit 55
- Compound 497
- Dieldrex
- Dielmoth
- Dildrin
- Dorytox
- Ent 16,225
- HEOD
- Hexachloroepoxyoctahydro-*endo,exo*-dimethanonaphthalene
- Illoxol
- Insectlack
- Insecticide no. 497
- Kombi-Albertan
- Moth snub D
- NCI-C00124
- Octalox
- Panoram D-31
- Quintox
- Red Shield
- SD 3417
- Termitox

CAS Registry No.: 60-57-1 (formerly 3039-00-7, 12622-75-2, 15113-81-2, 17301-10-9, 19237-11-7, 24502-07-6, 33648-22-5, 33737-28-9)
ROTECS Ref.: IO17500
Merck Index Ref.: 3075

Major Ions: 79, 82, 81, 77, 108, 53, 107, 263
EPA Ions: 79, 263, 279

Selected Bibliography

Applications of mass spectrometry to trace determinations of environmental toxic materials, Abramson, F. P., *Anal. Chem.*, **44**(14), 28A–33A,35A (1972).

Coupling of a liquid chromatograph to a mass spectrometer, Lovins, R. E., Ellis, S. R., Tolbert, G. D., McKinney, C. R., *Adv. Mass Spectrom.*, **6**, 457–462 (1974).

Isolation and characterization of some methanonaphthalene photoproducts, Onuska, F. I., Comba, M. E., *Biomed. Mass Spectrom.*, **2**(4), 169–175 (1975).

Identification of trace contaminants in environmental samples by selected ion summation analysis of gas chromatographic–mass spectral data, Kuehl, D. W., *Anal. Chem.*, **49**(3), 521–522 (1977).

Application of coupled gas chromatography–mass spectrometry in methods for the study and determination of pesticide residues and organic micropollutants in environmental and food materials, Mestres, R., Chevallier, C., Espinoza, C., Cornet, R., *Ann. Falsif. Exper. Chim.*, **70**(751), 177–188 (1977).

CHLORDANE (mixture)

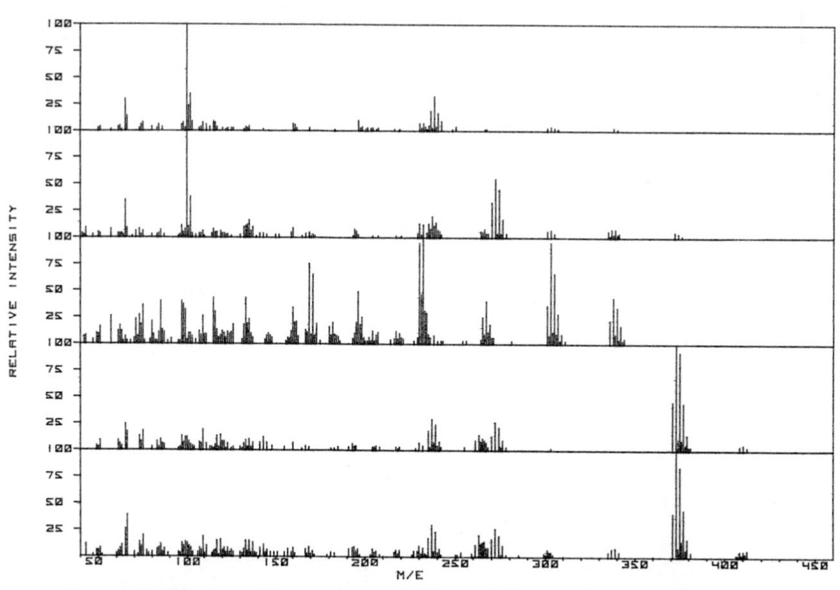

CHLORDANE–*continued*

Spectral Data

Peak 1

Mass	Abundance	Mass	Abundance	Mass	Abundance
50	3.0	113	3.7	206	2.2
51	4.4	115	9.3	207	3.0
57	2.2	116	8.1	216	2.2
61	3.7	117	4.1	219	2.2
62	4.8	120	2.6	230	7.0
63	2.2	122	2.2	231	2.6
65	30.4	123	2.6	232	6.7
66	14.4	125	2.6	233	3.7
73	3.3	126	3.3	234	2.2
74	6.3	132	2.2	235	5.2
75	7.8	133	4.4	236	19.3
80	3.7	134	3.3	237	3.3
83	3.3	135	4.8	238	33.0
84	5.6	143	2.2	239	4.4
86	4.1	160	6.7	240	17.4
97	5.9	161	5.6	242	9.3
98	8.1	162	3.3	248	2.2
99	3.0	169	3.0	250	3.7
100	100.0	183	2.2	266	2.2
101	23.7	196	10.4	267	2.2
102	35.2	197	2.6	301	2.6
103	8.9	198	4.4	303	3.7
107	2.6	200	2.2	305	3.3
108	4.1	201	3.0	307	2.2
109	7.8	203	3.3	338	3.0
111	6.3	204	2.6	340	2.2

Peak 2

Mass	Abundance	Mass	Abundance	Mass	Abundance
41	4.0	71	5.8	100	100.0
42	3.1	73	8.0	101	9.7
43	9.3	74	3.1	102	38.1
47	2.7	75	6.2	103	4.4
50	4.9	80	3.1	105	2.7
51	4.0	83	2.7	107	3.5
57	8.4	84	4.4	108	3.5
61	4.4	85	7.1	109	5.8
62	4.0	87	2.7	110	2.2
63	4.4	95	2.2	114	3.5
64	2.7	96	3.1	115	7.5
65	35.0	97	11.1	116	4.9
66	9.3	98	4.9	117	5.3
69	2.2	99	8.0	119	5.8

Peak 2–*continued*

Mass	Abundance	Mass	Abundance	Mass	Abundance
120	4.0	172	2.2	265	5.3
121	4.4	193	2.2	266	7.5
122	2.7	194	8.4	267	3.5
123	2.7	195	5.8	268	4.0
125	2.2	196	2.7	270	33.2
130	3.1	204	2.2	272	54.9
132	10.2	206	2.2	273	4.0
133	11.1	207	2.2	274	44.7
134	11.5	217	2.2	275	3.5
135	15.5	220	2.2	276	17.3
136	5.8	229	3.5	278	4.4
137	10.2	230	13.3	301	5.8
139	2.2	231	3.1	303	6.6
141	4.4	232	11.5	305	4.4
143	4.4	234	4.9	335	6.2
145	2.7	235	13.3	336	2.2
150	3.1	236	7.5	337	7.5
152	3.1	237	20.4	338	3.1
159	4.9	238	10.6	339	8.0
160	9.3	239	14.2	340	2.7
165	2.2	240	7.1	341	3.5
167	4.4	241	6.2	372	5.3
169	5.3	242	3.1	374	4.4
170	2.2	264	5.8	376	2.2
171	2.7				

Peak 3

Mass	Abundance	Mass	Abundance	Mass	Abundance
41	2.1	71	22.7	96	3.1
42	7.2	72	7.2	97	40.2
43	8.2	73	26.8	98	37.1
47	4.1	74	17.5	99	32.0
49	10.3	75	36.1	100	4.1
50	9.3	76	3.1	101	10.3
51	15.5	79	4.1	102	10.3
54	2.1	80	20.6	103	7.2
56	2.1	81	9.3	105	4.1
57	25.8	82	3.1	107	12.4
58	2.1	83	3.1	108	8.2
60	2.1	84	11.3	109	25.8
61	12.4	85	40.2	110	9.3
62	16.5	86	13.4	111	9.3
63	12.4	87	11.3	115	43.3
64	3.1	88	2.1	116	29.9
65	7.2	89	3.1	117	13.4
66	3.1	91	5.2	118	6.2
68	3.1	92	2.1	119	8.2
70	5.2	95	3.1	120	12.4

CHLORDANE–*continued*

Peak 3–*continued*

Mass	Abundance	Mass	Abundance	Mass	Abundance
121	10.3	185	7.2	252	2.1
122	6.2	186	3.1	253	2.1
123	11.3	187	2.1	254	3.1
124	9.3	193	5.2	255	2.1
125	11.3	194	10.3	256	3.1
126	17.5	195	19.6	258	2.1
131	4.1	196	48.5	264	4.1
132	17.5	197	17.5	265	24.7
133	43.3	198	24.7	266	9.3
134	18.6	199	6.2	267	40.2
135	22.7	200	3.1	268	11.3
136	10.3	201	3.1	269	17.5
137	6.2	202	6.2	270	6.2
143	2.1	203	3.1	271	6.2
144	4.1	204	12.4	281	3.1
145	8.2	205	3.1	282	2.1
146	10.3	206	8.2	284	2.1
147	8.2	207	11.3	289	2.1
148	6.2	210	2.1	300	2.1
150	2.1	211	2.1	301	36.1
151	2.1	213	2.1	302	8.2
154	2.1	214	4.1	303	94.8
156	3.1	216	5.2	304	11.3
157	6.2	217	12.4	305	66.0
158	5.2	218	5.2	306	9.3
159	12.4	219	10.3	307	27.8
160	34.0	220	6.2	308	3.1
161	19.6	221	5.2	309	9.3
162	20.6	227	3.1	310	2.1
163	7.2	228	2.1	311	3.1
164	2.1	229	6.2	325	2.1
167	13.4	230	94.8	327	2.1
168	11.3	231	46.4	336	21.6
169	75.3	232	100.0	338	43.3
170	9.3	233	30.9	339	9.3
171	64.9	234	28.9	340	34.0
172	8.2	235	9.3	341	4.1
173	18.6	236	6.2	342	16.5
175	3.1	238	8.2	343	3.1
180	15.5	240	3.1	344	5.2
181	8.2	241	2.1	346	2.1
182	19.6	242	3.1	374	2.1
183	9.3	243	3.1	376	2.1
184	8.2				

Peak 4

Mass	Abundance	Mass	Abundance	Mass	Abundance
49	4.3	120	8.2	230	6.5
50	2.6	121	8.2	232	3.9
51	8.6	122	2.2	235	18.1
61	8.6	123	5.6	236	2.6
62	5.6	125	2.2	237	28.9
63	4.3	126	3.0	238	6.9
65	23.7	130	2.6	239	24.1
66	16.8	131	2.2	240	4.3
73	12.5	132	5.6	241	8.2
74	7.8	133	9.1	242	3.0
75	18.1	134	3.4	255	2.6
80	2.6	135	9.9	261	9.1
83	8.2	136	2.6	263	14.2
84	3.4	137	6.9	264	8.2
85	9.5	141	7.3	265	10.8
86	5.6	143	12.1	266	8.6
87	4.7	145	7.3	267	7.3
95	2.2	147	4.3	268	3.4
96	3.4	155	2.6	270	13.4
97	12.5	160	7.3	272	26.3
98	7.3	165	2.2	274	20.7
99	12.1	167	3.9	275	3.0
100	11.6	169	3.0	276	9.1
101	7.8	181	2.2	278	3.0
102	4.7	183	2.2	303	2.2
103	3.9	185	2.6	371	44.8
104	2.6	191	3.4	373	100.0
107	6.9	193	5.6	374	10.3
108	5.6	194	3.9	375	91.8
109	19.0	195	3.9	376	9.1
111	6.0	204	3.4	377	44.4
113	3.0	205	3.4	378	4.7
114	3.4	206	4.3	379	13.8
115	2.2	208	3.0	380	2.6
116	4.7	217	2.6	381	3.0
117	12.9	218	2.2	408	4.3
118	2.6	219	3.4	410	4.7
119	13.8	228	2.2	412	3.0

Peak 5

Mass	Abundance	Mass	Abundance	Mass	Abundance
43	11.9	61	4.8	73	13.7
47	2.4	62	8.3	74	9.5
49	6.0	63	10.7	75	20.2
50	6.0	65	25.6	77	4.8
51	8.3	66	39.3	78	3.0
52	2.4	70	3.6	80	3.6
60	3.0	72	2.4	83	6.5

CHLORDANE–*continued*

Peak 5–*continued*

Mass	Abundance	Mass	Abundance	Mass	Abundance
84	7.7	145	6.0	241	7.1
85	11.9	147	4.2	242	2.4
86	4.2	149	4.2	250	2.4
87	7.1	151	3.6	251	2.4
89	3.0	155	4.2	253	4.2
95	4.2	157	6.5	259	3.0
96	3.0	159	4.2	261	11.3
97	12.5	160	7.7	263	20.2
98	9.5	161	3.0	264	11.9
99	14.3	167	6.5	265	13.1
100	12.5	168	3.0	266	13.7
101	9.5	169	8.9	267	8.3
102	7.7	170	3.0	268	8.3
103	4.2	171	4.8	270	16.1
106	3.6	172	2.4	272	25.6
107	8.3	179	2.4	274	19.0
108	6.0	181	3.6	275	2.4
109	19.0	183	3.0	276	9.5
110	3.0	191	7.1	278	2.4
111	10.1	193	7.7	285	2.4
114	3.0	194	8.9	299	3.0
115	5.4	195	4.8	300	2.4
116	6.0	196	7.1	301	6.0
117	14.9	197	2.4	302	4.2
118	2.4	203	2.4	303	3.6
119	16.1	204	7.1	304	2.4
120	6.0	205	3.6	335	4.2
121	7.7	206	5.4	337	6.5
122	3.6	208	2.4	339	7.7
123	7.7	216	3.6	341	3.6
124	3.6	217	6.0	371	40.5
125	6.0	218	3.0	373	100.0
126	4.2	219	6.0	374	8.3
130	3.6	226	2.4	375	83.9
131	3.0	227	4.8	376	13.7
132	7.1	229	4.8	377	43.5
133	14.9	230	10.1	378	6.0
134	4.8	231	3.0	379	15.5
135	14.9	232	7.7	381	4.2
136	3.6	233	3.0	406	2.4
137	13.1	234	2.4	407	2.4
138	4.2	235	16.7	408	4.8
139	3.0	237	29.8	409	2.4
141	8.3	238	5.4	410	4.8
143	10.7	239	22.6	411	3.0
144	3.6	240	3.6	412	6.0

Pesticide
CAS Name: Chlordane
Synonyms

Aspon	Corodane	Ortho-Klor
Belt	Ent 9,932	Prentox
CD68	Intox	Penicklor
Chlordan	Kypchlor	Synklor
γ-Chlordan	M 140	Tat chlor 4
Chlor-Kil	M 410	Topiclor
Chlorodane	Niran	Toxichlor
Chlorogran	Octachlor	Unexan-Koeder
Chlorodan	Octa-Klor	Velsicol 1068

CAS Registry No.: 12789-0306 (formerly 52002-35-4)
ROTECS Ref.: PB98000
Merck Index Ref.: 2051

Major Ions:
Peak 1: 100, 102, 238, 65, 101, 236, 240, 66
Peak 2: 100, 272, 274, 102, 65, 270, 237, 276
Peak 3: 232, 303, 230, 169, 305, 171, 196, 231
Peak 4: 373, 375, 371, 377, 237, 272, 239, 65
Peak 5: 373, 375, 377, 371, 66, 237, 272, 65

EPA Ions: 373, 375, 377

Selected Bibliography

Contributions to ecological chemistry, XCVI. On the structure of chlordene isomers of technical chlordane, Gaeb, S., Parlar, H., Cochrane, W. P., Fitzky, H. G., Wendisch, D., Korte, F., *Justus Liebigs Ann. Chem.*, (1), 1–12 (1976).

Structural elucidation of compound C. A chlordene isomer constituent of technical chlordane, Gaeb, S., Parlar, H., Korte, F., *J. Agric. Food Chem.*, **25**(5), 1224–1226 (1977).

PHENOL

Phenol is used as a disinfectant and in the manufacture of many industrial chemicals.

PHENOL C_6H_6O (94)

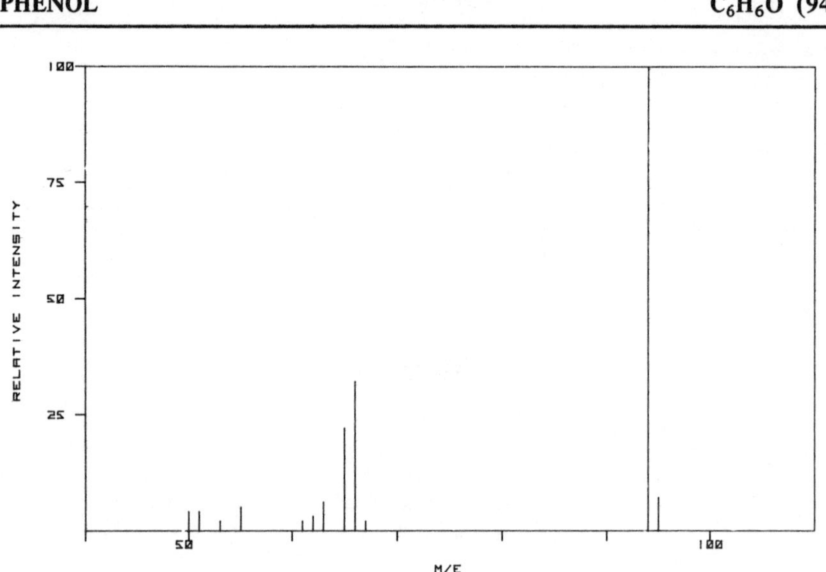

Spectral Data

Mass	Abundance	Mass	Abundance	Mass	Abundance
50	3.9	61	2.3	66	32.3
51	3.5	62	3.2	67	2.1
53	2.3	63	5.6	94	100.0
55	4.7	65	22.0	95	6.6

Acid extractable
CAS Name: Phenol
Synonyms

- Benzenol
- Carbolic acid
- Hydroxybenzene
- Izal
- Monohydroxybenzene
- Monophenol
- NCI-C50124
- Oxybenzene
- Phenic acid
- Phenyl alcohol
- Phenyl hydrate
- Phenyl hydroxide
- Phenylic acid
- Phenylic alcohol

PHENOL

CAS Registry No.: 108-95-2 (formerly 8002-07-1, 14534-23-7, 50356-25-7)
ROTECS Ref.: SJ33250
Merck Index Ref.: 7038

Major Ions: 94, 66, 65, 95, 63, 55, 50, 51
EPA Ions: 94, 65, 66

Selected Bibliography

Correlation between electronic fragmentation and the thermal decomposition of toluene and some phenols, Braekman-Danheux, C., Nguyen Cu Quyen, *Ann. Mines Belg.*, (2), 179–184 (1977).

A survey of the molecular nature of primary and second components of particles in urban air by high-resolution mass spectrometry, Cronn, D. R., Charlson, R. J., Knights, R. L., Crittenden, A. L., Appel, B. R., *Atmos. Environ.*, **11**(10), 929–937 (1977).

A gas chromatographic/mass spectrometric determination of some organic compounds of two Norwegian rocks, Langmyhr, F. J., Kolsaker, P., Steen, B. G., *Nor. Geol. Tidsskr.*, **57**(3), 285–294 (1977).

On the formation of polycyclic aromatics: investigation of fuel oil and emissions by high-resolution mass spectrometry, Herlan, A., *Combust. Flam.*, **31**(3), 297–307 (1978).

Characterization of liquids and gases obtained by hydrogenating lumps of Texas lignite, Philip, C. V., Anthony, R. G., *ACS Symp. Ser.*, **71**(Org. Chem. Coal), 258–273 (1978).

PHTHALATE ESTERS

Bis(2-ethylhexyl) phthalate, butyl benzyl phthalate, di-n-butyl phthalate, diethyl phthalate, dimethyl phthalate, and di-n-octyl phthalate are the compounds included in this category. Since bis(2-ethylhexyl) phthalate is frequently referred to as "dioctyl phthalate," care should be taken to distinguish it from di-n-octyl phthalate.

The phthalate esters are ubiquitous plasticizers, but some have other uses.

Bis(2-ethylhexyl) phthalate is used in vacuum pumps and was formerly employed as a miticide.

Di-n-butyl phthalate has been used as an insect repellant.

Diethyl phthalate has been used as a substitute for camphor in the manufacture of celluloid, as a solvent for cellulose acetate in the manufacture of varnishes and dopes, as a fixative for perfumes, and for denaturing ethanol.

Dimethyl phthalate is used as an insect repellant. It has also been employed as a solvent for cellulose acetate.

BIS(2-ETHYLHEXYL) PHTHALATE $C_{24}H_{38}O_4$ (390)

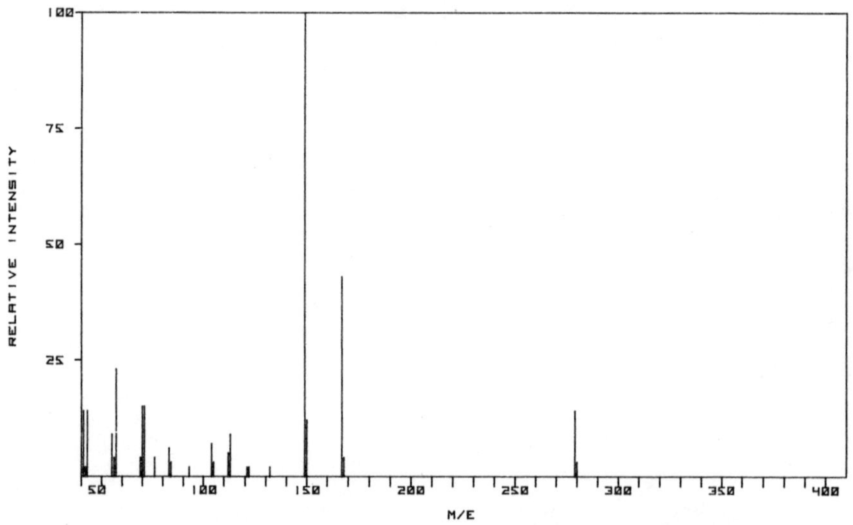

Spectral Data

Mass	Abundance	Mass	Abundance	Mass	Abundance
41	13.6	76	3.9	122	2.1
42	2.2	83	5.5	132	2.2
43	14.3	84	3.1	149	100.0
55	8.7	93	2.2	150	11.9
56	4.3	104	7.2	167	43.1
57	23.3	105	2.5	168	3.9
69	4.3	112	5.3	279	13.7
70	15.0	113	9.3	288	2.7
71	15.4	121	2.2		

Base–neutral extractable
CAS Name: Phthalic acid, bis(2-ethylhexyl) ester
Synonyms

Bis(2-ethylhexyl) 1,2-benzenedicarboxylate
Bis(2-ethylhexyl)-1,2-benzenedicarboxylic acid ester
Bisoflex 81
Compound 889
DEHP
Di(2-ethylhexyl)orthophthalate
Diethylhexyl phthalate
Di(ethylhexyl) phthalate
Di(2-ethylhexyl)phthalate
Di(*sec*-octyl) phthalate
DOP
Ethylhexyl phthalate
2-Ethylhexyl phthalate
Fleximel
Flexol plasticizer DOP
Hercoflex 260
Kodaflex DOP
NCI-C52733
Octoil
Octyl phthalate
Palatinol AH
Phthalic acid dioctyl ester
Pittsburgh PX-138
Sicol 150
Staflex DOP
Truflex DOP
Vestinol AH
Witcizer 312

CAS Registry No.: 117-81-7 (formerly 8033-53-2, 40120-69-2, 50885-87-5)
ROTECS Ref.: TI03500
Merck Index Ref.: 1270

Major Ions: 149, 167, 57, 71, 70, 43, 279, 41
EPA Ions: 149, 167, 279

Selected Bibliography

(Mass spectra of some phthalic acid derivatives), Cornell, R. R., Fenselau, C., *Arch. Mass Spectral Data.*, 2(4), 700–709 (1971).

Assessment of the trace organic molecular composition of industrial and municipal wastewater effluents by capillary gas chromatography/real-time high-resolution mass spectrometry: a preliminary report, Burlingame, A. L., *Ecotoxicol. Environ. Saf.*, 1(1), 111–150 (1977).

Application of coupled gas chromatography–mass spectrometry in methods for the study and determination of pesticide residues and organic micropollutants in environmental and food materials, Mestres, R., Chevallier, C., Espinoza, C., Cornet, R., *Ann. Falsif. Exper. Chim.*, 70(751), 177–188 (1977).

continued overleaf

BIS(2-ETHYLHEXYL) PHTHALATE–*continued*

Chromatographic profile of high boiling point organic acids in human urine, Brown, G. K., Stokke, O., Jellum, E., *J. Chromatogr.*, **145**(2), 177–184 (1978).

Determination of phthalic acid esters by gas chromatography-chemical ionization mass spectrometry, Kashiwagi, K., Iida, Y., Kokubun, N., Ikeda, T., *Seikei Daigaku Kogakubu Kogaku Hokoku*, **25**, 1779–1780 (1978).

BUTYL BENZYL PHTHALATE $C_{19}H_{20}O_4$ (312)

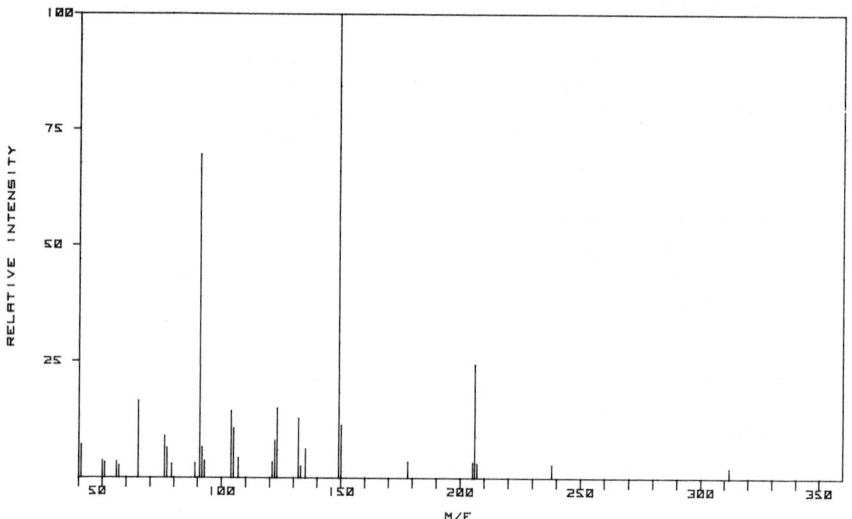

Spectral Data

Mass	Abundance	Mass	Abundance	Mass	Abundance
41	6.9	91	69.5	133	2.4
50	3.5	92	6.4	135	6.1
51	3.1	93	3.5	149	100.0
56	3.3	104	14.2	150	11.2
57	2.5	105	10.5	178	3.4
65	16.4	107	4.2	205	3.2
76	8.8	121	3.2	206	24.4
77	6.3	122	7.9	207	3.0
79	2.9	123	14.9	238	2.7
89	3.0	132	12.7	312	2.0

Base–neutral extractable
CAS Name: Phthalic acid, benzyl butyl ester
Synonyms

Benzyl butyl phthalate	Santicizer 160
Benzyl butyl phthalic acid ester	Sicol 160
NCI-C54375	Unimoll BB
Palatinol BB	

CAS Registry No.: 85-68-7 (formerly 58128-78-2)
ROTECS Ref.: TH99900

Major Ions: 149, 91, 206, 65, 123, 104, 132, 150
EPA Ions: 149, 91

Selected Bibliography

Application of coupled gas chromatography–mass spectrometry in methods for the study and determination of pesticide residues and organic micropollutants in environmental and food materials, Mestres, R., Chevallier, C., Espinoza, C., Cornet, R., *Ann. Falsif. Exper. Chim.*, **70**(751), 177–188 (1977).

Determination of phthalic acid esters by gas chromatography-chemical ionization mass spectrometry, Kashiwagi, T., Iida, Y., Kokubun, N., Ikeda, T., *Seikei Daigaku Kogakubu Kogaku Hokoku*, **25**, 1779–1780 (1978).

DI-n-BUTYL PHTHALATE $C_{16}H_{22}O_4$ (278)

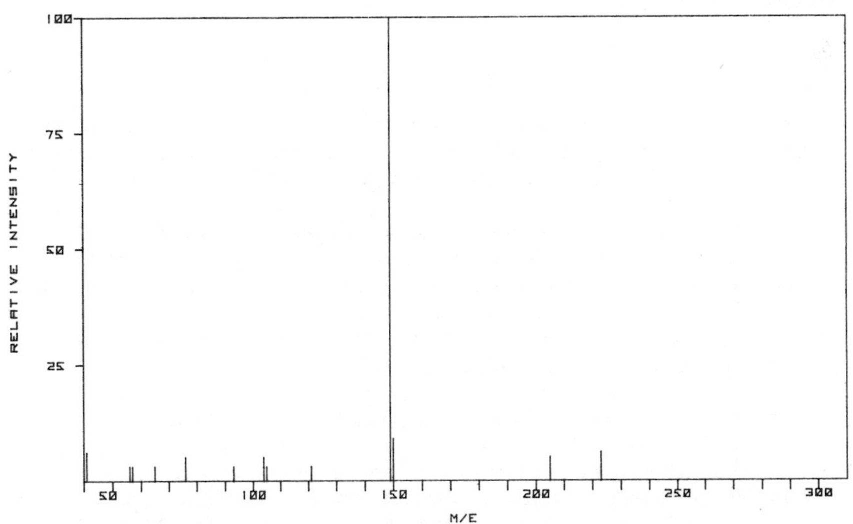

DI-n-BUTYL PHTHALATE–*continued*

Spectral Data

Mass	Abundance	Mass	Abundance	Mass	Abundance
41	6.3	93	3.3	149	100.0
56	2.9	104	5.0	150	9.0
57	2.7	105	2.6	205	4.5
65	3.4	121	2.5	223	5.7
76	4.8				

Base–neutral extractable
CAS Name: Phthalic acid, dibutyl ester

Synonyms

- *o*-Benzenedicarboxylic acid, dibutyl ester
- Benzene-*o*-dicarboxylic acid, di-n-butyl ester
- Butyl phthalate
- n-Butyl phthalate
- Celluflex DPB
- DBP
- Dibutyl phthalate
- Dibutyl-n-phthalate
- Dibutyl phthalic acid ester
- Elaol
- Genoplast B
- Hexaplas M/B
- Palatinol C
- Polycizer DBP
- PX 104
- Staflex DBP
- Unimoll DB
- Witcizer 300

CAS Registry No.: 84-74-2
ROTECS Ref.: TI08750
Merck Ref.: 1575

Major Ions: 149, 150, 41, 223, 104, 76, 205, 65
EPA Ions: 149, 150, 104

Selected Bibliography

Assessment of the trace organic molecular composition of industrial and municipal wastewater effluents by capillary gas chromatography/real-time high-resolution mass spectrometry: a preliminary report, Burlingame, A. L., *Ecotoxicol. Environ. Saf.*, **1**(1), 111–150 (1977).

Application of coupled gas chromatography–mass spectrometry in methods for the study and determination of pesticide residues and organic micropollutants in environmental and food materials, Mestres, R., Chevallier, C., Espinoza, C., Cornet, R., *Ann. Falsif. Exper. Chim.*, **70**(751), 177–188 (1977).

Concentration and analysis of trace impurities in styrene monomer, Zlatkis, A., Anderson, J. W., Holzer, G., *J. Chromatogr.*, **142**, 127–129 (1977).

Determination of phthalic acid esters by gas chromatography-chemical ionization mass spectrometry, Kashiwagi. T., Iida. Y., Kokubun, N., Ikeda, T., *Seikei Daigaku Kogakubu Kogaku Hokoku*, **25**, 1779–1780 (1978).

Feasibility of gunshot residue detection via its organic constitutents. Part I. Analysis of smokeless powders by combined gas chromatography–chemical ionization mass spectrometry, Mach, M. H., Pallos, A., Jones, P. F., *J. Forensic Sci.*, **23**(3), 433–445 (1978).

DIETHYL PHTHALATE $C_{12}H_{14}O_4$ (222)

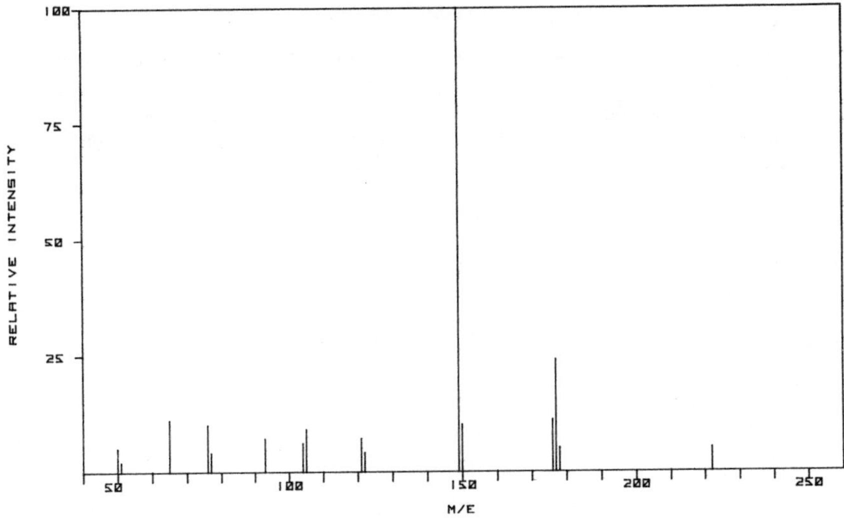

Spectral Data

Mass	Abundance	Mass	Abundance	Mass	Abundance
50	5.0	104	6.2	150	10.2
51	2.3	105	8.5	176	10.5
65	10.6	121	7.0	177	23.7
76	9.6	122	3.5	178	5.1
77	4.1	149	100.0	222	4.6
93	6.7				

Base–neutral extractable
CAS Name: Phthalic acid, diethyl ester
Synonyms

- Anozol
- 1,2-Benzenedicarboxylic acid, diethyl ester
- *o*-Benzenedicarboxylic acid, diethyl ester
- Diethyl phthalic acid ester
- Ethyl phthalate
- Neantine
- Palatinol A
- Phthalol
- Placidol E
- Solvanol
- Unimoll DA

CAS Registry No.: 84-66-2
ROTECS Ref.: TI10500

Major Ions: 149, 177, 65. 176, 150, 76, 105, 121
EPA Ions: 149, 177, 150

DIETHYL PHTHALATE–*continued*

Selected Bibliography

Concentration and analysis of trace impurities in styrene monomer, Zlatkis, A., Anderson, J. W., Holzer, G., *J. Chromatogr.*, **142**, 127–129 (1977).

Determination of phthalic acid esters by gas chromatography-chemical ionization mass spectrometry, Kashiwagi, T., Iida, Y., Kokubun, N., Ikeda, T., *Seikei Daigaku Kogakubu Kogaku Hokoku*, **25**, 1779–1780 (1978).

DIMETHYL PHTHALATE $C_{10}H_{10}O_4$ (194)

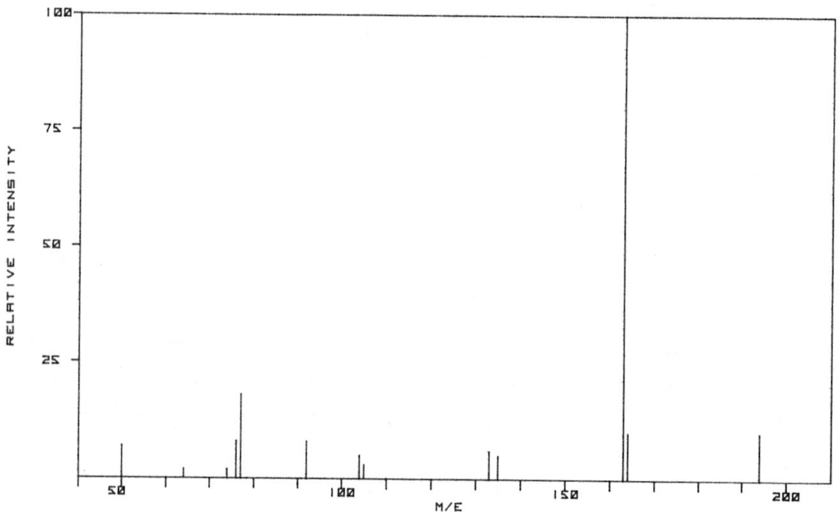

Spectral Data

Mass	Abundance	Mass	Abundance	Mass	Abundance
50	7.2	92	8.0	135	5.3
64	2.1	104	4.6	163	100.0
74	2.1	105	3.1	164	9.5
76	7.6	133	6.2	194	10.2
77	17.7				

Base–neutral extractable

CAS Name: Phthalic acid, dimethyl ester

Synonyms

- Avolin
- 1,2-Benzenedicarboxylic acid, dimethyl ester
- Dimethyl 1,2-benzenedicarboxylate
- Dimethyl phthalic acid ester
- DMP
- Ent 262

Synonyms–*continued*

Fermine
Methyl phthalate
Mipax
NTM
Palatinol M
Phthalic acid methyl ester
Solvanom
Solvarone
Unimoll DM

CAS Registry No.: 131-11-3
ROTECS Ref.: TI15750
Merck Index Ref.: 3244

Major Ions: 163, 77, 194, 164, 92, 76, 50, 133
EPA Ions: 163, 194, 164

Selected Bibliography

Rapid method for the mass spectrometric identification of glucuronides and other polar drug metabolites in permethylated rat bile, Thompson, R. M., Gerber, N., Seibert, R. A., Desiderio, D. M., *Drug Metab. Dispos.*, **1**(2), 489–505 (1973).

Application of coupled gas chromatography–mass spectrometry in methods for the study and determination of pesticide residues and organic micropollutants in environmental and food materials, Mestres, R., Chevallier, C., Espinoza, C., Cornet, R., *Ann. Falsif. Exper. Chim.*, **70**(751), 177–188 (1977).

Concentration and analysis of trace impurities in styrene monomer, Zlatkis, A., Anderson, J. W., Holzer, G., *J. Chromatogr.*, **142**, 127–129 (1977).

Real-time analysis of gaseous atmospheric pollutants to the ppt level using a mobile API mass spectrometer system, Reid, N. M., French, J. B., Buckley, J. A., Lane, D. A., Lovett, A. M., Rosenblatt, G., *4th Jt. Conf. Sens. Environ. Pollut.*, American Chemical Society, Washington, D.C., (1978), 594–600.

DI-n-OCTYL PHTHALATE $C_{24}H_{38}O_4$ (390)

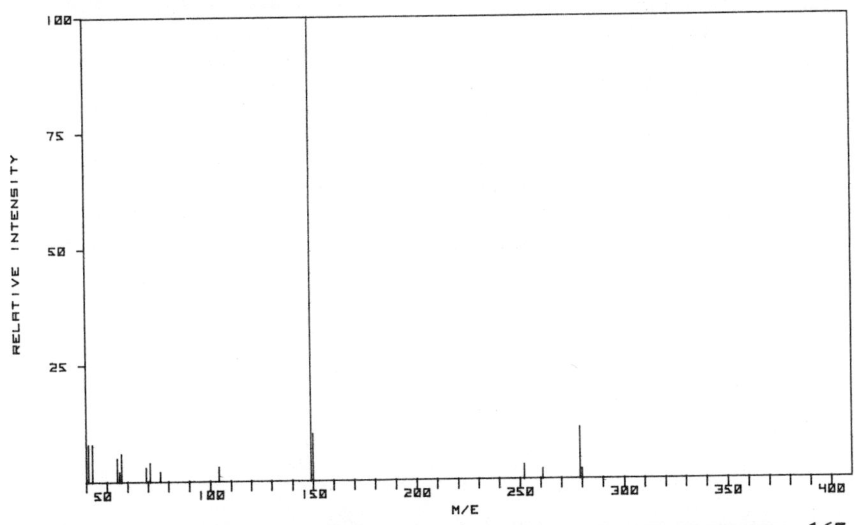

DI-n-OCTYL PHTHALATE–*continued*

Spectral Data

Mass	Abundance	Mass	Abundance	Mass	Abundance
41	7.5	69	3.2	150	9.7
43	7.6	71	3.8	252	2.5
55	4.5	76	2.3	261	2.2
56	2.3	104	2.5	279	10.7
57	5.5	149	100.0	280	2.2

Base–neutral extractable
CAS Name: Phthalic acid, dioctyl ester
Synonyms

- *o*-Benzenedicarboxylic acid, dioctyl ester
- Celluflex DOP
- Diethyl hexyl phthalate
- Dinopol NOP
- Dioctyl *o*-benzenedicarboxylate
- Dioctyl phthalate
- n-Dioctyl phthalate
- Dioctyl phthalic acid ester
- Octyl phthalate
- n-Octyl phthalate
- Polycizer 162
- PX-138

CAS Registry No.: 117-84-0 (formerly 8031-29-6, 14374-99-3)
ROTECS Ref.: TI19250

Major Ions: 149, 279, 150, 43, 41, 57, 71, 69
EPA Ion: 149

Selected Bibliography

Analysis of high-boiling monomer plasticizers. Identification of trimellitic acid esters, Gross, D., Metasch, W., *Kaut. Gummi, Kunstst.*, **24**(10), 532–535 (1971).

Phthalic acid esters in rock bituminoids as an index of high metamorphism of dispersed organic matter, Klindukhov, V. P., Surgova, N. Z., Glebovskaya, E. A., Mel'tsanskaya, T. N., Kulikova, E. M., Solov'eva, I. L., Belyaeva, L. S., Bikkenina, D. A., Sarbeeva, L. I., *Tr. Vses. Neft. Nauchno-Issled. Geologorazved. Inst.*, **353**, 58–64 (1974).

Some examples of measurement by chemical ionization, Yamauchi, E., *Bunseki Kiki*, **14**(4), 197–206 (1976).

Application of coupled gas chromatography–mass spectrometry in methods for the study and determination of pesticide residues and organic micropollutants in environmental and food materials, Mestres, R., Chevallier, C., Espinoza, C., Cornet, R., *Ann. Falsif. Exper. Chim.*, **70**(751), 177–188 (1977).

POLYCHLORINATED BIPHENYLS

These are mixtures of compounds containing differing numbers of chlorine atoms per molecule. The mixtures included this category are PCB-1016, 1221, 1232, 1242, 1248, 1254, and 1260. The last two digits refer to the degree of chlorination: thus, PCB-1016 contains 16% chlorine and PCB-1260 contains 60% chlorine. These designations correspond to the Monsanto designations Aroclor 1016, etc. ("Aroclor" is frequently misspelled as "Arochlor"). Aroclors 5432, 5442, and 5460 are polychlorinated terphenyls. PCBs are marketed in Japan as Kanechlors and in France as Prodelecs.

PCBs were formerly in widespread use as plasticizers, lubricants, transformer fluids, etc., but their use is now severely restricted in the U.S.

PCBs are degraded by dechlorination, so the GC profiles of weathered heavier mixtures resemble those of the lighter mixtures. Care should therefore be exercised in ascribing an origin to a PCB mixture.

PCB-1016 **(mixture)**

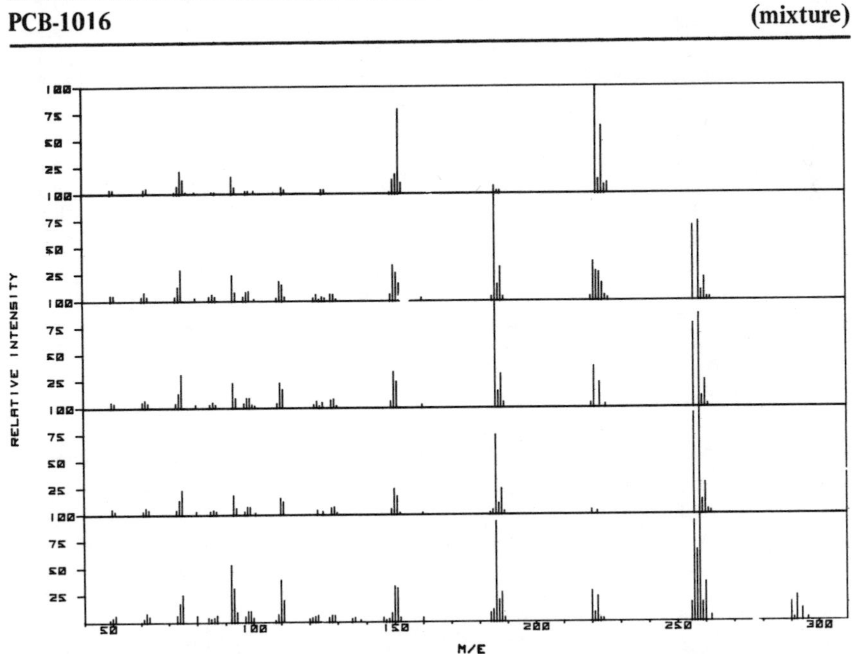

PCB-1016–*continued*

Spectral Data

Peak 1

Mass	Abundance	Mass	Abundance	Mass	Abundance
50	3.9	93	15.5	151	17.6
51	3.6	94	5.8	152	78.7
62	4.1	98	3.2	153	9.5
63	5.4	99	3.3	186	6.9
73	2.1	101	2.7	187	2.5
74	6.6	111	6.2	188	2.5
75	20.9	112	3.7	222	100.0
76	12.9	125	3.5	223	13.1
77	2.0	126	4.4	224	62.9
80	2.4	149	2.3	225	8.4
86	2.3	150	13.2	226	10.0
87	2.2				

Peak 2

Mass	Abundance	Mass	Abundance	Mass	Abundance
50	5.3	109	2.7	186	100.0
51	4.9	110	18.1	187	14.6
61	3.9	111	15.1	188	30.9
62	7.6	112	4.3	189	4.3
63	4.4	122	2.7	220	3.5
73	4.0	123	5.6	221	36.1
74	13.4	124	2.1	222	26.9
75	29.1	125	4.3	223	25.8
80	3.3	126	2.9	224	16.0
85	3.9	128	6.1	225	5.2
86	5.5	139	6.1	226	2.5
87	3.7	130	2.2	256	69.3
93	24.3	149	6.5	258	72.6
94	7.8	150	32.7	259	8.8
97	3.8	151	25.6	260	21.3
98	8.4	152	16.4	261	3.3
99	8.5	160	2.6	262	2.5
101	2.3	185	3.9		

Peak 3

Mass	Abundance	Mass	Abundance	Mass	Abundance
50	4.9	80	2.6	99	9.0
51	4.4	85	3.0	100	2.5
61	4.6	86	5.1	101	2.1
62	6.7	87	3.4	109	3.9
63	3.9	93	23.3	110	23.4
73	3.6	94	8.5	111	17.2
74	13.4	97	4.1	122	2.6
75	30.8	98	9.2	123	6.1

Peak 3–*continued*

Mass	Abundance	Mass	Abundance	Mass	Abundance
124	2.3	160	2.6	223	22.8
125	4.8	186	100.0	225	3.3
128	7.4	187	15.4	256	78.4
129	7.5	188	31.0	258	87.4
130	2.1	189	4.6	259	10.2
149	5.7	220	3.6	260	24.9
150	33.1	221	38.0	261	2.8
151	23.9				

Peak 4

Mass	Abundance	Mass	Abundance	Mass	Abundance
50	5.0	98	6.7	184	2.0
51	3.4	99	6.9	185	3.9
61	3.1	101	2.2	186	73.6
62	5.8	110	15.2	187	10.3
63	3.5	111	12.0	188	23.7
73	3.8	123	4.3	189	2.7
74	13.0	125	3.2	220	4.4
75	22.6	128	5.5	221	2.6
80	2.6	129	6.5	256	94.1
85	2.6	130	2.0	258	100.0
86	4.0	149	5.0	259	12.5
87	2.6	150	24.3	260	28.6
93	17.9	151	17.0	261	3.9
94	6.0	152	2.4	262	3.3
97	3.0	160	2.4		

Peak 5

Mass	Abundance	Mass	Abundance	Mass	Abundance
49	2.2	94	9.4	129	6.3
50	4.0	97	5.4	135	3.1
51	5.8	98	10.3	136	4.0
61	2.7	99	10.3	138	2.2
62	8.0	100	4.0	146	3.6
63	4.5	108	2.2	147	2.2
73	6.3	109	6.7	148	3.1
74	16.5	110	38.8	149	8.0
75	25.4	111	20.1	150	33.0
80	5.8	120	2.7	151	30.8
84	3.6	121	3.6	152	4.0
85	3.1	122	4.9	160	4.0
86	4.0	123	6.3	184	7.6
87	5.8	124	2.2	185	10.7
92	52.7	127	3.6	186	92.9
93	31.3	128	5.8	187	20.1

PCB-1016—*continued*

Peak 5—*continued*

Mass	Abundance	Mass	Abundance	Mass	Abundance
188	27.2	225	16.5	262	5.4
189	4.0	256	92.9	290	17.4
220	28.1	257	66.1	291	2.7
221	8.0	258	100.0	292	23.2
222	22.8	259	17.0	294	11.2
223	3.1	260	36.2	296	2.7
224	3.1				

Pesticide
CAS Name: Aroclor 1016
CAS Registry No.: 12674-11-2

Major Ions
- Peak 1: 222, 152, 224, 75, 151, 93, 150, 223
- Peak 2: 186, 258, 256, 221, 150, 188, 75, 222
- Peak 3: 186, 258, 256, 221, 150, 75, 260, 151
- Peak 4: 258, 256, 186, 260, 150, 188, 75, 93
- Peak 5: 258, 256, 186, 257, 92, 110, 260, 150

EPA Ions: (not listed)

PCB-1221 (mixture)

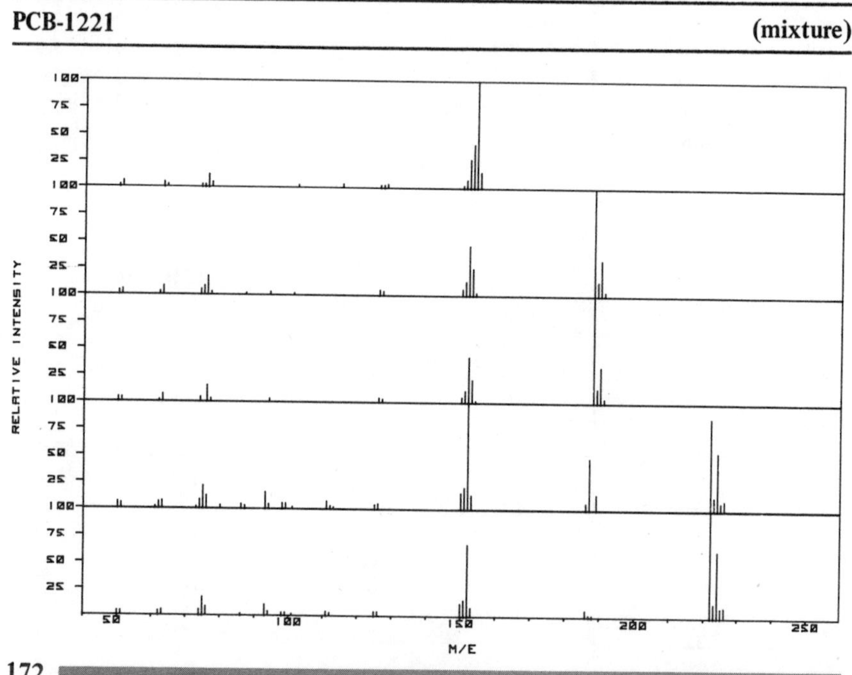

Spectral Data

Peak 1

Mass	Abundance	Mass	Abundance	Mass	Abundance
50	3.3	77	5.4	150	2.6
51	5.5	102	2.7	151	7.5
63	5.0	115	4.1	152	26.9
64	2.5	126	3.1	153	40.8
74	2.9	127	2.6	154	100.0
75	3.3	128	3.8	155	14.5
76	11.6				

Peak 2

Mass	Abundance	Mass	Abundance	Mass	Abundance
50	3.8	87	2.2	152	45.9
51	4.8	94	2.7	153	25.0
62	2.7	101	2.1	154	2.7
63	8.0	126	5.1	188	100.0
74	4.9	127	3.5	189	13.3
75	8.3	150	5.6	190	32.8
76	16.5	151	13.3	191	4.2
77	3.2				

Peak 3

Mass	Abundance	Mass	Abundance	Mass	Abundance
50	3.6	77	2.8	153	20.6
51	4.2	94	3.4	154	2.3
62	2.2	126	3.8	188	100.0
63	6.8	127	3.4	189	13.2
74	3.7	150	4.7	190	32.5
75	6.5	151	11.2	191	3.7
76	14.7	152	41.8		

Peak 4

Mass	Abundance	Mass	Abundance	Mass	Abundance
50	5.5	93	15.3	151	20.0
51	5.0	94	4.3	152	100.0
61	2.4	98	4.5	153	12.5
62	5.8	99	5.2	186	6.2
63	6.6	101	2.4	187	48.4
73	2.0	111	6.5	189	14.2
74	8.3	112	3.3	222	86.1
75	20.8	113	2.0	223	11.9
76	11.5	125	4.3	224	54.3
80	2.7	126	5.3	225	6.7
86	4.0	150	14.9	226	9.0
87	3.1				

PCB-1221–*continued*

Peak 5

Mass	Abundance	Mass	Abundance	Mass	Abundance
50	4.0	98	2.8	153	7.9
51	3.7	99	3.3	186	5.8
62	4.0	101	2.3	187	2.4
63	5.1	111	4.4	188	2.2
74	5.4	112	2.8	222	100.0
75	16.9	125	3.5	223	12.8
76	8.4	126	3.9	224	62.1
86	2.2	150	11.9	225	8.5
93	10.2	151	14.8	226	10.3
94	3.6	152	67.2		

Pesticide
CAS Name: Aroclor 1221
CAS Registry No.: 11104-28-2
ROTECS Ref.: TQ13520

Major Ions Peak 1: 154, 153, 152, 155, 76, 151, 51, 77
Peak 2: 188, 152, 190, 153, 76, 189, 151, 75
Peak 3: 188, 152, 190, 153, 76, 189, 151, 63
Peak 4: 152, 222, 224, 187, 75, 151, 93, 150
Peak 5: 222, 152, 224, 75, 151, 223, 150, 226

EPA Ions: (not listed)

PCB-1232 (mixture)

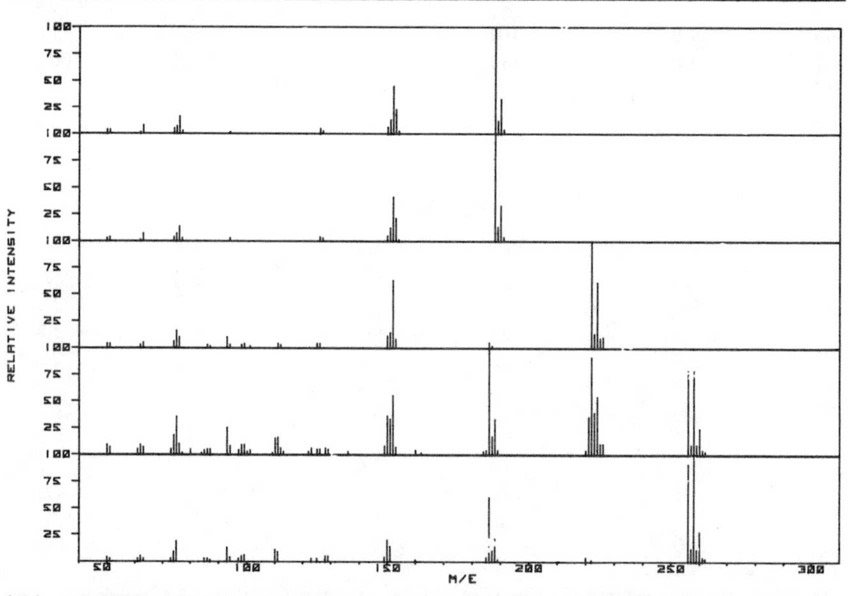

Spectral Data

Peak 1

Mass	Abundance	Mass	Abundance	Mass	Abundance
50	3.6	77	3.2	153	23.0
51	4.3	94	2.4	154	2.7
62	2.4	126	4.7	188	100.0
63	7.8	127	3.4	189	12.2
74	4.7	150	5.7	190	32.5
75	6.8	151	12.8	191	3.9
76	15.6	152	44.8		

Peak 2

Mass	Abundance	Mass	Abundance	Mass	Abundance
50	3.2	77	2.9	153	20.6
51	4.2	94	3.3	154	2.3
62	2.1	126	3.7	188	100.0
63	6.8	127	3.3	189	12.5
74	3.7	150	4.6	190	32.7
75	6.6	151	11.7	191	4.1
76	13.7	152	41.4		

Peak 3

Mass	Abundance	Mass	Abundance	Mass	Abundance
50	3.7	94	3.3	152	63.4
51	3.9	98	2.9	153	7.6
62	3.4	99	3.7	186	5.4
63	4.9	101	2.3	187	2.4
74	5.6	111	4.4	222	100.0
75	16.4	112	2.8	223	13.1
76	9.5	125	3.6	224	60.9
86	2.5	126	3.9	225	8.7
87	2.1	150	10.9	226	9.5
93	10.1	151	13.9		

Peak 4

Mass	Abundance	Mass	Abundance	Mass	Abundance
50	8.5	80	4.8	100	2.5
51	6.8	84	2.3	101	3.8
61	5.1	85	3.6	109	2.0
62	8.7	86	5.1	110	15.0
63	7.1	87	4.5	111	15.9
73	5.2	93	24.6	112	5.5
74	18.2	94	7.9	113	2.7
75	35.4	97	3.9	122	2.5
76	10.3	98	9.1	123	5.6
77	2.0	99	9.0	125	5.2

PCB-1232–*continued*

Peak 4–*continued*

Mass	Abundance	Mass	Abundance	Mass	Abundance
126	4.8	184	2.6	224	53.5
128	5.8	185	3.8	225	10.3
129	4.8	186	100.0	226	9.5
136	2.6	187	16.8	256	78.9
149	7.7	188	33.2	257	9.2
150	35.7	189	3.9	258	79.0
151	32.8	220	3.6	259	9.4
152	54.9	221	34.8	260	23.6
153	6.8	222	91.2	261	4.0
160	3.6	223	38.7	262	2.9
162	2.2				

Peak 5

Mass	Abundance	Mass	Abundance	Mass	Abundance
50	4.0	98	4.8	186	60.1
51	2.8	99	5.5	187	9.6
61	2.9	110	11.4	188	21.0
62	5.0	111	9.4	189	2.2
63	2.5	123	3.3	220	4.2
73	2.5	125	2.9	222	2.3
74	9.1	128	4.8	256	91.1
75	19.3	129	4.8	257	11.8
85	3.1	149	4.2	258	100.0
86	3.2	150	20.3	259	11.3
87	2.4	151	14.4	260	27.7
93	13.3	160	2.1	261	3.9
94	4.0	185	3.5	262	3.0
97	2.9				

Pesticide
CAS Name: Aroclor 1232
CAS Registry No.: 11141-16-5
ROTECS Ref.: TQ13540
Major Ions
Peak 1: 188, 152, 190, 153, 76, 151, 189, 63
Peak 2: 188, 152, 190, 153, 76, 189, 151, 63
Peak 3: 222, 152, 224, 75, 151, 223, 150, 93
Peak 4: 186, 222, 258, 256, 152, 224, 223, 150
Peak 5: 258, 256, 186, 260, 188, 150, 75, 151
EPA Ions: (not listed)

PCB-1242 (mixture)

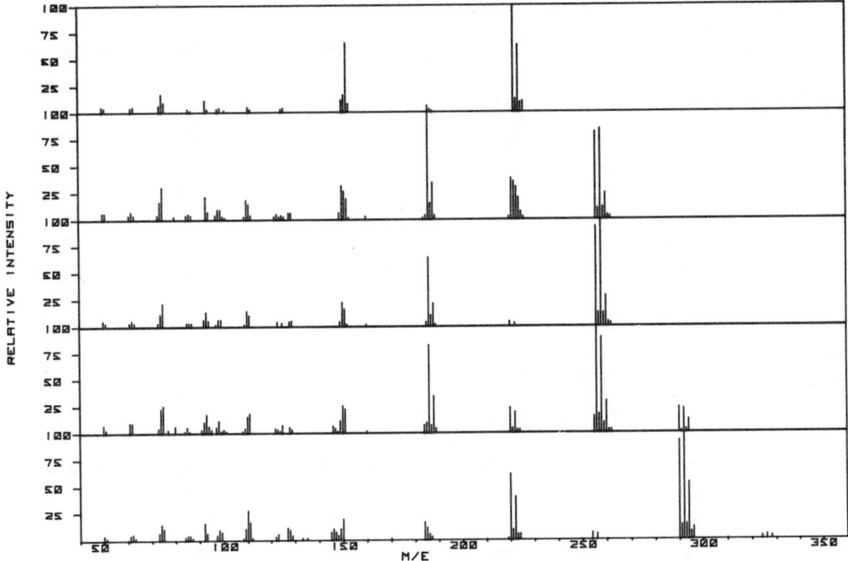

Spectral Data

Peak 1

Mass	Abundance	Mass	Abundance	Mass	Abundance
50	4.7	94	3.4	152	65.0
51	4.2	98	3.3	153	8.1
62	3.9	99	3.6	186	6.3
63	5.3	101	2.3	187	2.6
74	6.3	111	4.6	188	2.3
75	16.6	112	2.5	222	100.0
76	9.1	125	3.3	223	12.4
86	2.8	126	4.2	224	62.5
87	2.4	150	11.4	225	8.7
93	11.3	151	16.0	226	9.7

Peak 2

Mass	Abundance	Mass	Abundance	Mass	Abundance
50	6.3	75	29.7	97	3.7
51	6.3	80	2.8	98	9.0
61	4.3	85	4.0	99	8.9
62	7.3	86	4.7	100	2.6
63	4.4	87	4.2	101	2.3
73	4.3	93	21.4	109	2.6
74	15.5	94	6.9	110	17.7

PCB-1242–*continued*

Peak 2–*continued*

Mass	Abundance	Mass	Abundance	Mass	Abundance
111	13.8	152	18.8	223	29.6
112	3.6	153	2.0	224	20.3
122	3.1	160	3.0	225	6.7
123	5.2	184	2.3	226	3.3
124	2.7	185	4.1	256	80.7
125	4.3	186	100.0	257	10.4
126	2.6	187	14.5	258	84.4
128	6.4	188	34.2	259	10.6
129	6.4	189	3.8	260	24.0
149	6.4	220	3.4	261	4.0
150	31.4	221	37.6	262	2.9
151	25.5	222	35.1		

Peak 3

Mass	Abundance	Mass	Abundance	Mass	Abundance
50	4.5	98	5.7	185	4.0
51	3.2	99	6.0	186	63.9
61	3.3	109	2.3	187	9.8
62	5.1	110	13.9	188	20.8
63	3.4	111	9.5	189	2.4
73	3.0	123	3.7	220	4.0
74	10.8	125	2.9	222	2.7
75	20.8	128	4.4	256	92.9
85	2.5	129	4.9	257	12.1
86	3.3	149	4.2	258	100.0
87	2.5	150	21.7	259	11.5
92	6.2	151	15.7	260	28.0
93	13.2	152	2.1	261	4.4
94	5.1	160	2.0	262	3.3
97	2.4				

Peak 4

Mass	Abundance	Mass	Abundance	Mass	Abundance
50	7.1	85	4.9	100	2.7
51	2.7	86	2.2	101	2.2
61	8.8	91	2.7	108	2.2
62	9.3	92	9.9	109	4.4
73	4.4	93	16.5	110	15.4
74	22.0	94	6.0	111	17.6
75	24.7	95	2.7	122	4.4
77	2.7	97	4.9	123	3.3
80	5.5	98	11.0	124	2.2
84	2.2	99	2.2	125	7.1

Peak 4–*continued*

Mass	Abundance	Mass	Abundance	Mass	Abundance
128	4.9	186	81.9	257	17.0
129	2.7	187	7.1	258	89.0
146	5.5	188	33.5	259	9.3
147	4.4	189	4.4	260	29.1
148	2.2	220	22.5	261	2.7
149	11.0	221	4.4	262	3.3
150	25.3	222	19.2	290	22.5
151	22.0	223	3.3	291	2.2
160	2.2	224	2.7	292	22.0
184	6.6	255	15.4	293	2.7
185	8.8	256	100.0	294	11.5

Peak 5

Mass	Abundance	Mass	Abundance	Mass	Abundance
50	4.4	111	15.6	187	3.1
51	2.3	112	2.2	219	2.4
61	4.4	122	3.2	220	61.0
62	4.6	123	15.0	221	9.4
63	2.0	127	10.5	222	40.2
73	5.5	128	9.0	223	5.0
74	13.8	129	3.8	224	5.2
75	10.2	133	2.3	254	5.9
84	2.7	135	2.0	256	4.7
85	4.3	145	6.6	290	92.2
86	3.8	146	9.7	291	12.9
87	2.4	147	7.4	292	100.0
92	14.8	148	3.5	293	14.4
93	5.6	149	9.7	294	53.1
97	4.3	150	19.1	295	7.0
98	8.5	170	2.3	296	10.5
99	7.1	184	15.5	324	3.2
108	2.3	185	10.6	326	4.3
109	9.5	186	4.8	328	3.4
110	27.3				

Pesticide
CAS Name: Aroclor 1242
CAS Registry No.: 53469-21-9 (formerly 11104-29-3)
ROTECS Ref.: TQ13560

Major Ions
Peak 1: 222, 152, 224, 75, 151, 223, 150, 93
Peak 2: 186, 258, 256, 221, 222, 188, 150, 75
Peak 3: 258, 256, 186, 260, 150, 188, 75, 151
Peak 4: 256, 258, 186, 188, 260, 150, 75, 290
Peak 5: 292, 290, 220, 294, 222, 110, 150, 111
EPA Ions: 224, 260, 294

PCB-1248 (mixture)

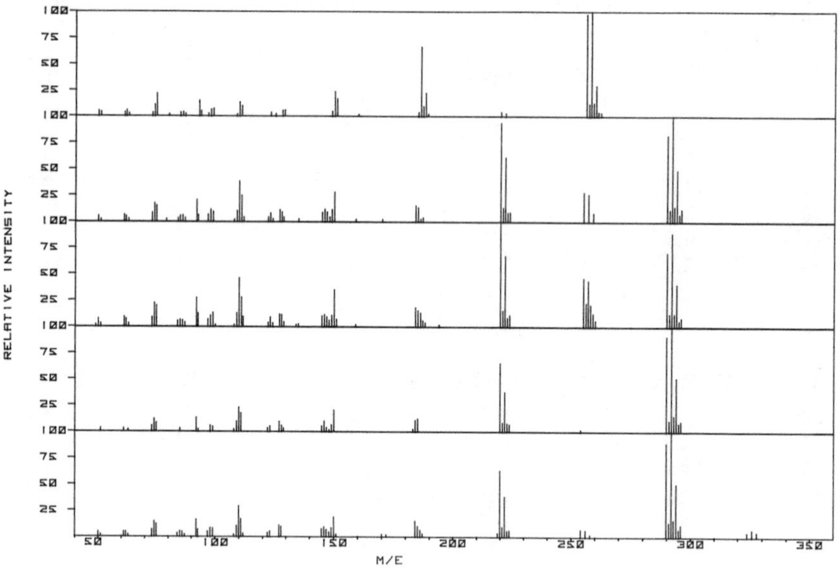

Spectral Data

Peak 1

Mass	Abundance	Mass	Abundance	Mass	Abundance
50	5.0	97	2.7	185	3.7
51	4.1	98	6.1	186	66.3
61	3.7	99	7.0	187	9.5
62	5.5	109	2.2	188	22.2
63	3.0	110	13.5	189	2.6
73	3.5	111	9.6	220	4.1
74	10.6	123	3.6	222	3.1
75	21.5	125	2.6	256	97.0
80	2.1	128	5.2	257	11.9
85	3.4	129	5.5	258	100.0
86	3.9	149	4.4	259	12.9
87	2.7	150	23.6	260	29.4
93	14.9	151	16.7	261	4.0
94	4.7	160	2.1	262	3.4

Peak 2

Mass	Abundance	Mass	Abundance	Mass	Abundance
50	5.3	109	10.1	184	15.1
51	2.4	110	37.9	185	13.4
61	6.3	111	24.5	186	2.7
62	5.0	112	4.3	187	3.8
63	2.8	122	4.4	220	93.4
73	8.3	123	7.8	221	13.2
74	17.2	124	2.8	222	60.7
75	15.0	127	11.0	223	8.3
79	2.6	128	8.8	224	8.7
84	3.3	129	4.3	255	27.8
85	5.3	135	2.7	257	25.8
86	5.7	145	8.5	259	8.0
87	3.6	146	11.8	290	81.5
92	20.4	147	8.8	291	10.8
93	6.0	148	4.4	292	100.0
97	6.4	149	11.4	293	14.1
98	11.3	150	27.9	294	48.6
99	9.4	159	2.6	295	6.6
108	2.4	170	2.6	296	11.5

Peak 3

Mass	Abundance	Mass	Abundance	Mass	Abundance
49	2.4	111	27.7	187	5.6
50	7.4	112	9.3	188	4.0
51	3.2	122	3.7	194	2.1
61	9.0	123	8.8	220	100.0
62	7.2	124	3.5	221	14.9
63	3.2	127	11.4	222	67.0
73	8.8	128	11.2	223	8.2
74	22.1	129	4.5	224	10.6
75	19.7	134	2.1	255	45.7
84	5.1	135	2.7	256	22.1
85	6.1	145	10.1	257	43.6
86	5.6	146	11.2	258	20.5
87	4.0	147	9.0	259	12.2
92	26.9	148	5.9	260	5.6
93	12.5	149	10.6	290	70.2
97	6.6	150	34.8	291	11.7
98	10.4	151	6.9	292	88.8
99	13.0	159	2.1	293	11.7
100	2.1	184	18.1	294	39.9
108	2.1	185	15.2	295	5.1
109	12.8	186	13.0	296	8.0
110	45.7				

PCB-1248–*continued*

Peak 4

Mass	Abundance	Mass	Abundance	Mass	Abundance
51	3.4	111	17.2	185	12.0
61	3.1	122	3.4	220	64.9
63	2.2	123	4.9	221	8.3
73	5.2	127	9.2	222	37.2
74	11.7	128	5.5	223	7.4
75	8.3	129	3.4	224	6.8
85	3.1	145	4.9	254	2.2
92	12.9	146	9.5	290	90.5
93	2.5	147	3.7	291	10.2
98	5.5	148	2.2	292	100.0
99	4.6	149	6.2	293	14.5
108	2.5	150	20.0	294	50.8
109	9.5	183	2.8	295	7.4
110	22.5	184	10.8	296	9.5

Peak 5

Mass	Abundance	Mass	Abundance	Mass	Abundance
50	4.5	111	16.7	219	3.4
51	2.2	112	3.1	220	62.9
61	4.5	122	3.6	221	9.0
62	4.5	123	5.0	222	38.3
63	2.1	127	10.7	223	5.3
73	6.4	128	9.5	224	5.8
74	14.1	145	7.1	254	6.7
75	11.7	146	9.0	256	6.0
84	3.1	147	6.5	258	2.3
85	5.0	148	4.2	290	88.8
86	4.6	149	8.1	291	13.4
87	2.3	150	18.6	292	100.0
92	15.6	151	2.4	293	15.8
93	6.6	170	2.3	294	50.0
97	4.7	172	2.0	295	6.4
98	8.0	184	14.8	296	11.2
99	7.9	185	9.8	324	4.0
108	2.1	186	6.1	326	6.5
109	10.2	187	2.9	328	4.4
110	29.1				

Pesticide
CAS Name: Aroclor 1248
CAS Registry No.: 12672-29-6
ROTECS Ref.: TQ13580

Major Ions
Peak 1: 258, 256, 186, 260, 150, 188, 75, 151
Peak 2: 292, 220, 290, 222, 294, 110, 150, 255
Peak 3: 220, 292, 290, 222, 255, 110, 257, 294
Peak 4: 292, 290, 220, 294, 222, 110, 150, 111
Peak 5: 292, 290, 220, 294, 222, 110, 150, 111
EPA Ions: (not listed)

PCB-1254 (mixture)

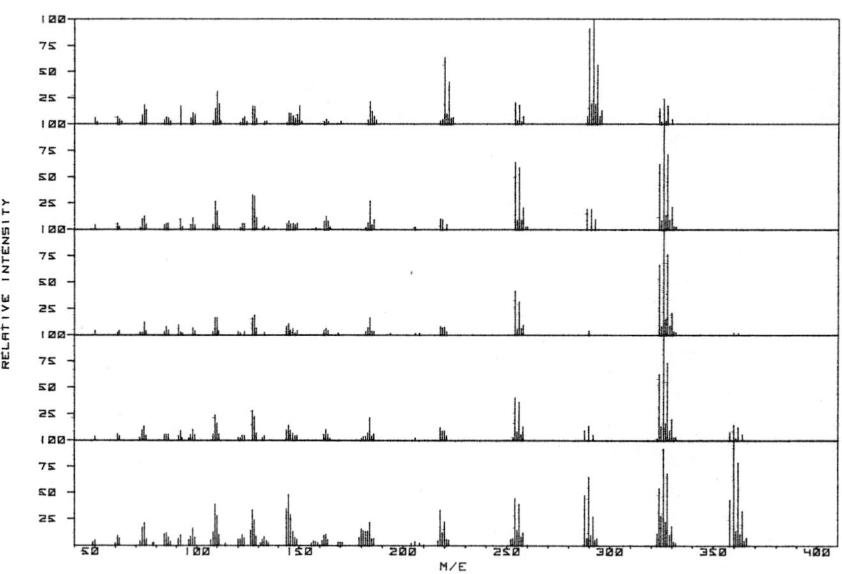

Spectral Data

Peak 1

Mass	Abundance	Mass	Abundance	Mass	Abundance
50	5.9	75	13.5	99	9.0
51	2.2	84	4.1	108	3.2
61	6.9	85	6.3	109	14.5
62	4.8	86	5.6	110	31.3
63	2.9	87	3.1	111	19.0
72	2.1	92	17.1	112	3.1
73	8.4	97	5.8	121	2.1
74	18.3	98	10.7	122	5.2

PCB-1254–*continued*

Peak 1–*continued*

Mass	Abundance	Mass	Abundance	Mass	Abundance
123	6.9	164	2.5	257	2.5
124	2.8	170	2.8	258	7.2
127	17.0	183	4.1	289	7.2
128	16.7	184	21.2	290	90.9
129	5.3	185	11.9	291	19.4
133	2.7	186	7.3	292	100.0
134	2.8	187	3.9	293	18.8
144	2.1	218	3.1	294	56.2
145	10.4	219	4.5	295	7.3
146	10.0	220	63.6	296	12.9
147	7.7	221	9.7	324	14.9
148	5.3	222	40.2	325	2.4
149	9.0	223	5.5	326	23.7
150	17.4	224	6.0	327	3.1
151	2.8	254	20.2	328	17.4
162	2.7	255	3.4	330	4.6
163	4.4	256	18.0		

Peak 2

Mass	Abundance	Mass	Abundance	Mass	Abundance
50	4.3	128	31.6	218	9.7
61	5.8	129	11.0	219	9.0
62	3.0	132	2.0	221	4.7
72	2.2	133	3.5	254	63.8
73	9.7	135	2.2	255	8.7
74	12.4	144	5.7	256	58.8
75	5.0	145	8.0	257	8.5
84	4.3	146	5.2	258	20.5
85	5.5	147	5.5	259	2.5
86	6.0	148	4.3	260	2.5
92	9.8	149	5.7	289	19.0
93	2.8	158	2.0	291	19.2
97	4.5	162	7.7	293	9.5
98	10.9	163	12.5	324	62.1
99	4.8	164	7.7	325	8.5
108	4.7	165	2.5	326	100.0
109	26.5	182	2.2	327	13.2
110	17.2	183	6.2	328	71.0
111	3.5	184	26.7	329	8.8
121	2.2	185	4.3	330	20.9
122	5.3	186	9.2	331	2.8
123	5.3	206	2.5	332	2.8
127	33.1				

Peak 3

Mass	Abundance	Mass	Abundance	Mass	Abundance
50	4.2	127	16.1	218	7.9
61	2.7	128	18.8	219	6.7
62	4.2	129	6.4	220	7.3
72	2.4	133	2.7	221	3.3
73	2.4	144	7.9	254	41.5
74	12.1	145	10.6	255	5.5
75	3.6	146	4.5	256	31.2
84	3.0	147	6.1	257	6.1
85	7.9	148	2.1	258	9.4
86	4.5	149	4.2	290	3.9
91	9.4	162	4.2	324	66.4
92	2.7	163	6.1	325	8.2
93	2.1	164	4.2	326	100.0
98	6.7	169	2.1	327	14.5
99	4.2	182	3.3	328	76.4
108	3.3	183	6.7	329	8.5
109	16.1	184	16.7	330	20.9
110	16.1	185	3.6	331	3.3
111	3.9	186	3.6	332	2.7
120	3.3	194	2.1	360	2.4
121	2.1	206	2.1	362	2.1
123	3.6	208	2.1		

Peak 4

Mass	Abundance	Mass	Abundance	Mass	Abundance
50	3.5	111	5.5	180	2.2
61	5.6	120	2.6	181	3.3
62	4.0	121	2.3	182	3.5
72	2.8	122	4.5	183	6.7
73	9.3	123	4.3	184	21.2
74	12.7	127	27.9	185	4.0
75	4.4	128	22.0	186	5.5
84	5.1	129	6.5	206	2.4
85	5.3	132	2.5	218	11.8
86	5.1	133	4.2	219	8.5
91	3.9	144	9.4	220	8.8
92	8.9	145	13.9	221	4.3
93	2.4	146	9.2	253	2.8
96	2.4	147	6.5	254	40.2
97	5.0	148	4.2	255	7.7
98	9.8	149	4.5	256	35.8
99	4.9	162	5.3	257	5.1
108	5.5	163	9.9	258	12.5
109	23.4	164	5.3	288	9.1
110	15.8	165	2.1	290	13.1

PCB-1254–*continued*

Peak 4–*continued*

Mass	Abundance	Mass	Abundance	Mass	Abundance
292	4.7	328	73.1	358	7.0
323	2.2	329	8.6	360	14.1
324	62.3	330	19.6	361	2.2
325	12.8	331	2.6	362	11.9
326	100.0	332	3.2	364	5.4
327	15.8				

Peak 5

Mass	Abundance	Mass	Abundance	Mass	Abundance
49	2.9	132	5.4	220	22.1
50	4.7	133	7.8	221	5.4
60	2.2	134	4.4	222	4.9
61	8.8	135	2.7	252	5.6
62	6.4	144	34.3	253	6.1
72	3.7	145	47.8	254	44.4
73	16.4	146	28.9	255	14.2
74	20.8	147	12.5	256	39.0
75	5.6	148	7.1	257	7.8
78	2.2	149	5.6	258	11.8
79	2.7	156	2.2	288	47.3
84	10.3	157	3.9	289	6.6
85	11.5	158	3.4	290	64.5
86	7.6	159	2.7	291	9.3
87	3.4	161	4.2	292	27.0
91	6.4	162	9.6	293	4.9
92	9.6	163	10.3	294	6.4
96	4.9	164	5.1	323	10.8
97	8.1	169	2.9	324	54.2
98	15.9	170	2.7	325	27.9
99	7.4	171	2.7	326	91.7
107	4.9	179	6.9	327	21.8
108	12.3	180	15.2	328	68.1
109	38.7	181	13.5	329	9.8
110	27.5	182	12.7	330	17.9
111	9.8	183	13.2	331	3.4
114	2.2	184	21.3	332	2.5
120	5.6	185	5.9	358	42.9
121	5.6	186	6.1	360	100.0
122	9.8	204	2.2	361	13.2
123	6.6	206	3.7	362	78.7
126	14.0	208	2.2	363	10.5
127	33.3	217	4.2	364	32.4
128	23.8	218	33.1	365	4.2
129	8.6	219	11.8	366	7.1
131	2.2				

Pesticide
CAS Name: Aroclor 1254
CAS Registry No.: 11097-69-1
ROTECS Ref.: TQ13600

Major Ions Peak 1: 292, 290, 220, 294, 222, 110, 326, 184
Peak 2: 326, 328, 254, 324, 256, 127, 128, 184
Peak 3: 326, 328, 324, 254, 256, 330, 128, 184
Peak 4: 326, 328, 324, 254, 256, 127, 128, 109
Peak 5: 360, 326, 362, 328, 324, 145, 288, 254
EPA Ions: 294, 330, 362

PCB-1260 (mixture)

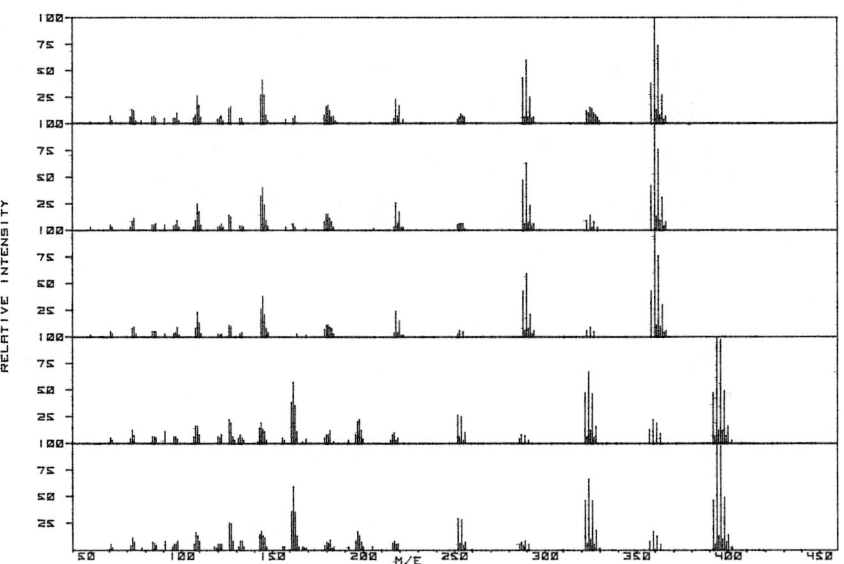

Spectral Data

Peak 1

Mass	Abundance	Mass	Abundance	Mass	Abundance
50	2.2	85	7.3	109	26.0
61	7.0	86	5.3	110	17.0
62	3.1	91	5.0	111	5.9
72	6.1	96	5.3	120	3.9
73	13.4	97	4.7	121	5.3
74	12.0	98	9.5	122	7.3
75	3.1	99	3.4	123	2.8
78	3.4	107	4.7	126	13.7
84	6.4	108	8.1	127	15.9

PCB-1260—*continued*

Peak 1—*continued*

Mass	Abundance	Mass	Abundance	Mass	Abundance
132	5.0	204	2.2	294	5.6
133	4.5	216	2.2	323	12.3
134	2.2	217	4.7	324	9.8
144	26.5	218	22.9	325	15.1
145	41.3	219	7.3	326	13.7
146	27.4	220	16.5	327	9.8
147	7.8	222	3.9	328	8.1
148	3.4	252	3.6	329	5.6
158	3.9	253	6.4	330	3.4
162	5.3	254	8.9	358	37.7
163	6.7	255	6.7	360	100.0
179	8.1	256	6.4	361	12.6
180	15.9	288	42.5	362	73.5
181	16.8	289	6.4	363	8.4
182	12.0	290	60.3	364	27.1
183	6.4	291	5.9	365	3.6
184	6.7	292	25.1	366	7.0
185	2.5	293	4.5		

Peak 2

Mass	Abundance	Mass	Abundance	Mass	Abundance
50	2.7	132	4.1	252	5.0
61	5.2	133	3.9	253	5.9
62	3.1	134	2.5	254	5.5
72	2.8	144	32.2	255	6.1
73	7.7	145	39.7	256	2.2
74	10.5	146	24.0	288	46.5
84	5.3	147	8.6	289	5.8
85	5.4	148	3.5	290	62.7
86	5.7	158	2.5	291	7.3
91	5.0	162	5.9	292	23.3
96	4.4	163	3.2	293	3.7
97	5.4	169	2.1	294	6.3
98	8.5	179	7.6	323	9.2
99	2.8	180	14.5	325	13.7
107	3.4	181	15.0	326	2.8
108	8.9	182	11.1	327	7.8
109	24.8	183	8.1	329	3.1
110	17.0	184	3.3	358	42.4
111	4.7	206	2.0	360	100.0
120	3.1	217	3.3	361	12.5
121	4.3	218	25.9	362	76.4
122	5.8	219	5.7	363	9.3
123	2.7	220	17.2	364	30.6
126	13.7	221	2.5	365	4.1
127	12.0	222	2.6	366	8.0

Peak 3

Mass	Abundance	Mass	Abundance	Mass	Abundance
50	2.2	127	9.7	222	2.4
61	5.3	132	3.1	252	2.7
62	2.7	133	3.8	253	5.5
73	7.7	144	26.1	255	5.1
74	8.8	145	37.8	288	43.1
75	2.6	146	20.5	289	5.7
84	4.9	147	7.5	290	59.4
85	4.8	148	4.2	291	7.9
86	5.3	164	2.7	292	20.5
91	2.9	165	2.4	293	2.9
96	3.3	169	2.0	294	5.5
97	4.0	179	6.8	323	6.0
98	9.3	180	11.2	325	9.0
99	2.4	181	11.3	327	5.1
108	7.5	182	9.1	358	42.8
109	22.9	183	7.7	360	100.0
110	13.3	184	3.1	361	11.2
111	2.9	217	3.5	362	75.5
120	2.7	218	24.3	363	9.7
121	2.4	219	4.4	364	30.3
122	3.3	220	15.0	365	4.0
126	11.3	221	2.4	366	6.4

Peak 4

Mass	Abundance	Mass	Abundance	Mass	Abundance
61	4.5	129	3.0	181	8.3
62	3.4	131	3.8	182	12.0
72	3.8	132	8.3	184	2.3
73	12.0	133	5.3	192	3.0
74	6.8	134	3.4	196	7.9
84	6.4	142	2.3	197	19.5
85	5.6	143	13.9	198	21.8
86	4.5	144	18.8	199	11.7
91	10.9	145	13.2	200	4.1
96	6.4	146	11.3	215	3.4
97	6.4	147	2.6	216	7.9
98	3.8	156	4.5	217	10.2
107	6.4	157	2.6	218	3.0
108	15.8	161	38.0	219	4.9
109	16.2	162	57.1	252	26.3
110	8.3	163	35.0	253	5.6
120	5.6	164	10.5	254	25.2
121	4.9	165	4.5	255	3.0
122	7.5	167	2.3	256	9.8
126	21.8	169	4.1	286	3.8
127	18.8	179	5.3	287	7.5
128	6.0	180	7.5	289	6.8

PCB-1260–*continued*

Peak 4–*continued*

Mass	Abundance	Mass	Abundance	Mass	Abundance
291	2.6	357	13.2	395	12.0
322	47.0	359	22.2	396	96.6
323	5.6	361	18.8	397	12.0
324	66.5	363	9.4	398	49.2
325	11.7	392	46.6	399	6.8
326	45.5	393	7.1	400	15.8
327	5.3	394	100.0	402	3.4
328	16.2				

Peak 5

Mass	Abundance	Mass	Abundance	Mass	Abundance
61	4.5	145	12.8	253	6.1
62	2.1	146	10.9	254	28.1
72	4.3	147	3.1	255	4.0
73	11.1	156	3.3	256	7.3
74	7.3	157	2.6	286	3.5
78	2.4	161	35.5	287	6.9
84	6.6	162	59.3	288	3.5
85	6.1	163	35.0	289	8.0
86	4.0	164	12.5	291	4.7
91	7.6	165	3.1	322	46.3
95	2.6	167	2.6	323	5.7
96	5.4	168	2.4	324	66.2
97	4.7	169	2.1	325	10.2
98	8.0	179	3.5	326	45.6
107	5.4	180	7.1	327	5.0
108	16.3	181	6.4	328	17.7
109	12.8	182	8.7	330	2.1
110	7.6	183	2.1	357	8.3
118	2.8	184	3.1	359	17.3
119	2.1	192	2.8	361	12.8
120	5.4	196	8.0	363	5.4
121	5.0	197	16.8	392	46.1
122	5.0	198	13.0	393	4.7
126	24.8	199	7.3	394	100.0
127	23.9	200	3.1	395	13.0
128	7.6	205	2.6	396	96.7
131	2.6	216	6.1	397	9.7
132	8.3	217	8.3	398	49.2
133	7.6	218	4.5	399	6.9
134	3.3	219	4.7	400	14.2
143	14.4	252	28.6	402	2.8
144	16.8				

Pesticide
CAS Name: Aroclor 1260
CAS Registry No.: 11096-82-5
ROTECS Ref.: TQ13620

Major Ions
- Peak 1: 360, 362, 290, 288, 145, 358, 146, 364
- Peak 2: 360, 362, 290, 288, 358, 145, 144, 364
- Peak 3: 360, 362, 290, 288, 358, 145, 364, 144
- Peak 4: 394, 396, 324, 162, 398, 322, 392, 326
- Peak 5: 394, 396, 324, 162, 398, 322, 392, 326

EPA Ions: (not listed)

Selected Bibliography

Analysis of organochlorine metabolites in crude extracts by high resolution photoplate mass spectrometry, Safe, S., Platonow, N., Hutzinger, O., Jamieson, W. D., *Biomed. Mass Spectrom.*, 2(4), 201–203 (1975).

The finding of polychlorodibenzofurans in commercial PCBs (Aroclor, Phenoclor and Clophen), Miyata, H., Kashimoto, T., *Shokuhin Eiseigaku Zasshi*, 17(6), 434–437 (1976).

POLYNUCLEAR AROMATIC HYDROCARBONS

Acenaphthylene, anthracene, benzo[*a*]anthrancene, benzo[*b*]fluoranthene, benzo[*k*] fluoranthene, benzo[*ghi*] perylene, benzo[*a*] pyrene, chrysene, dibenzo-[*a*,*h*]anthracene, fluorene, indeno[1,2,3-*cd*]pyrene, phenanthrene, and pyrene are the compounds included in this category.

These compounds are present in coal and petroleum products and are formed during the combustion of many materials. A major problem in their analysis is distinguishing between isomers, since their mass spectra and retention times (on most stationary phases) are almost identical. Some success, however, has been reported when liquid crystal phases are employed.

Acenaphthylene should not be confused with acenaphthene (*q.v.*).

Anthracene is used in dyestuff manufacture. It is also a major constituent of coal tar neutral oils, used with soap to control screw worms.

ACENAPHTHYLENE $C_{12}H_8$ (152)

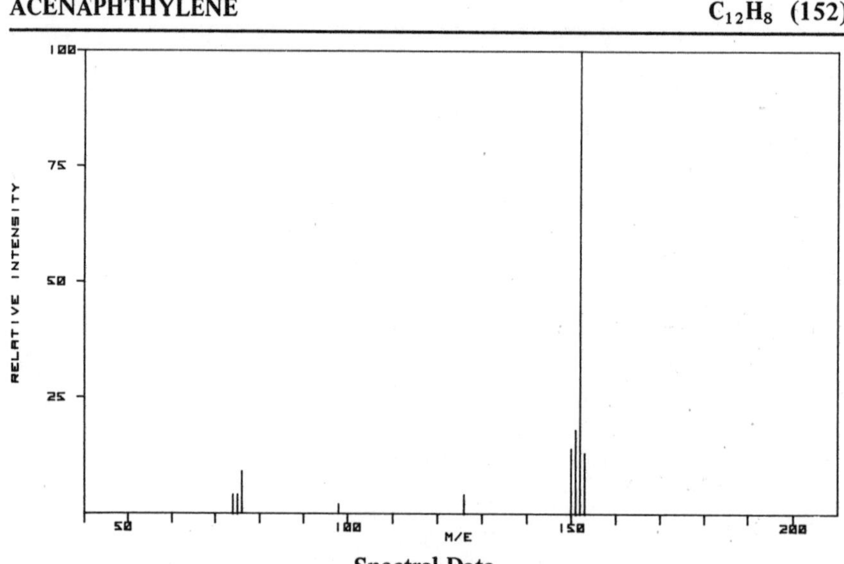

Spectral Data

Mass	Abundance	Mass	Abundance	Mass	Abundance
74	3.7	98	2.3	151	17.5
75	3.8	126	3.6	152	100.0
76	9.3	150	13.5	153	12.6

Base–neutral extractable
CAS Name: Acenaphthylene
Synonym:
Cyclopenta[*de*]naphthalene

CAS Registry No.: 208-96-8

Major Ions: 152, 151, 150, 153, 76, 75, 74, 126
EPA Ions: 152, 151, 153

Selected Bibliography

Electron impact studies of some cyclic hydrocarbons, Frey, W. F., Compton, R. N., Naff, W. T., Schweinler, H. C., *Int. J. Mass Spectrom. Ion Phys.*, **12**(1), 19–32 (1973).

A system for the rapid identification of toxic organic pollutants in water, Donaldson, W. T., Carter, M. H., McGuire, J. M., Comm. Eur. Communities, (Rep.) EUR. 1975, (EUR 5360, Proc. Int. Symp. Recent Advances Assessment Health Effects Environmental Pollutants, Vol. 3.), 1399–1406.

Analysis of complex polycyclic aromatic hydrocarbon mixtures by computerized GC/MS. Hites, R. A., *Prepr., Div. Pet. Chem., Am. Chem. Soc.*, **20**(4), 824–828 (1975).

Characterization of sulfur-containing polycyclic aromatic compounds in carbon blacks, Lee, M. L., Hites, R. A., *Anal. Chem.*, **48**(13), 1890–1893 (1976).

Mixed charge exchange–chemical ionization mass spectrometry of polycyclic aromatic hydrocarbons, Lee, M. L., Hites, R. A., *J. Am. Chem. Soc.*, **99**(6), 2008–2009 (1977).

ANTHRACENE $C_{14}H_{10}$ (178)

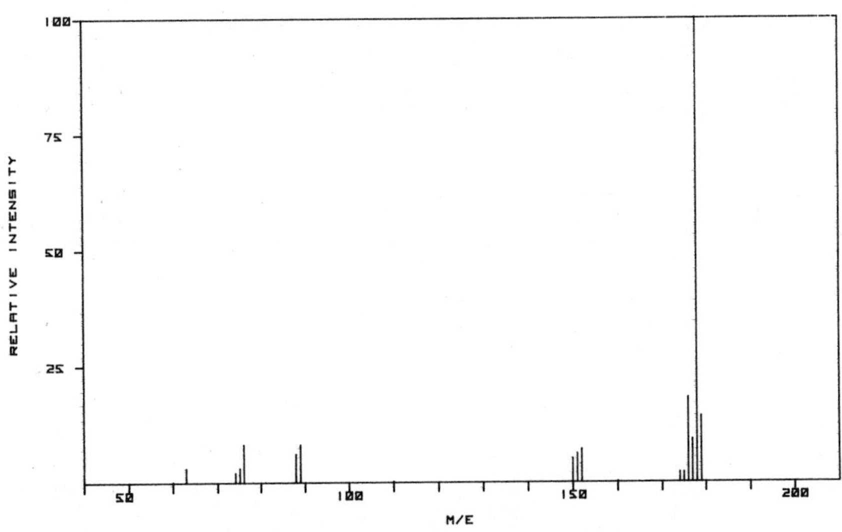

ANTHRACENE–*continued*

Spectral Data

Mass	Abundance	Mass	Abundance	Mass	Abundance
63	2.6	89	7.6	175	2.4
74	2.2	150	4.6	176	17.7
75	3.4	151	5.6	177	8.6
76	8.2	152	7.2	178	100.0
88	5.5	174	2.2	179	13.9

Base–neutral extractable
CAS Name: Anthracene
Synonyms
- Anthracin
- Green oil
- Paranaphthalene
- Tetra olive N2G

CAS Registry No.: 120-12-7 (formerly 4820-00-2)
ROTECS Ref.: CA93500
Merck Index Ref.: 718

Major Ions: 178, 176, 179, 177, 76, 89, 152, 151
EPA Ions: 178, 179, 176

Selected Bibliography

Doubly-charged ion mass spectra of hydrocarbons, Ast, T., Beynon, J. H., Cooks, R. G., *Org. Mass Spectrom.*, **6**(7), 749–763 (1972).
Multiply charged ions in the mass spectra of aromatics, Engel, R., Halpern, D., Funk, B. A., *Org. Mass Spectrom.*, **7**(2), 177–183 (1973).
Use of low-voltage mass spectrometry for analyzing coal tar refining products, Kekin, N. A., Shmal'ko, V. M., Vodolazhchenko, V. V., *Zavod. Lab.*, **43**(5), 551–553 (1977).
Determination of polynuclear aromatic hydrocarbons contaminated with chlorinated hydrocarbon pesticides, Negishi, T., *Bull. Environ. Contam. Toxicol.*, **19**(5), 545–548 (1978).

BENZO[*a*]ANTHRACENE $C_{18}H_{12}$ (228)

Spectral Data

Mass	Abundance	Mass	Abundance	Mass	Abundance
63	2.1	111	2.1	224	6.3
75	3.0	112	10.0	225	4.9
87	3.5	113	23.0	226	26.5
88	5.8	114	22.8	227	8.1
99	3.2	200	3.9	228	100.0
100	9.5	201	2.1	229	20.6
101	12.5	202	2.6		

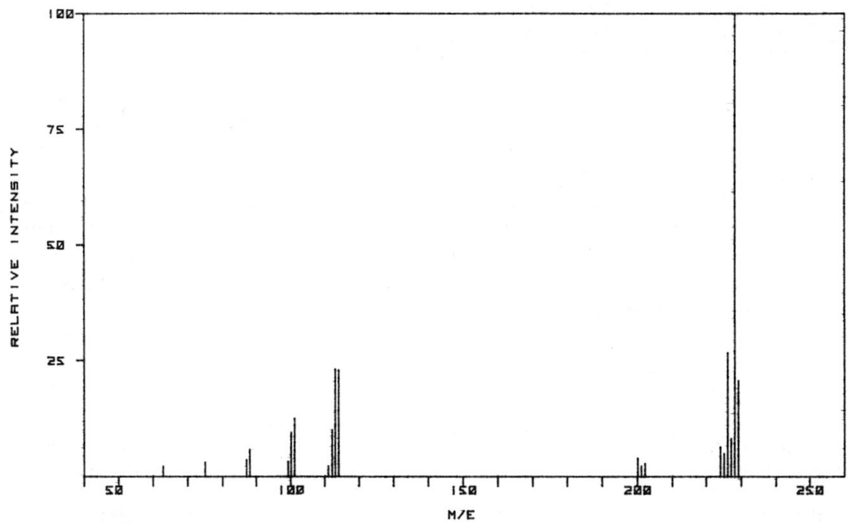

Base–neutral extractable
CAS Name: Benz[*a*]anthracene
Synonyms

Benzanthracene	Benzoanthracene	2,3-Benzophenanthrene
1,2-Benzanthracene	1,2-Benzoanthracene	2,3-Benzphenanthrene
1,2-Benz[*a*]anthracene	Benzo[*a*]phenanthrene	Naphthanthracene
Benzanthrene	Benzo[*b*]phenanthrene	Tetraphene
1,2-Benzanthrene		

CAS Registry No.: 56-55-3
ROTECS Ref.: CV92750
Merck Index Ref.: 1063

Major Ions: 228, 226, 113, 114, 229, 101, 112, 100
EPA Ions: 228, 229, 226

Selected Bibliography

Gas–liquid chromatographic evaluation and gas-chromatography/mass spectrometric application of new high-temperature liquid crystal stationary phases for polycylic aromatic hydrocarbon separation, Janini, G. M., Muschik, G. M., Schroer, J. A., Zielinsk, W. L., Jr., *Anal. Chem.*, 48(13), 1879–1883 (1976).

Characterization of sulfur-containing polycyclic aromatic compounds in carbon blacks, Lee, M. L., Hites, R. A., *Anal. Chem.*, 48(13), 1890–1893 (1976).

Mixed charge exchange–chemical ionization mass spectrometry of polycyclic aromatic hydrocarbons, Lee, M. L., Hites, R. A., *J. Am. Chem. Soc.*, 99(6), 2008–2009 (1977).

Gas chromatography–mass spectrometry of simulated arson residue using gasoline as an accelerant, Mach, M. H., *J. Forensic Sci.*, 22(2), 348–357 (1977).

continued overleaf

BENZO[*a*]ANTHRACENE–*continued*

A comparison of some chromatographic methods for estimation of polynuclear aromatic hydrocarbons in pollutants, Burchill, P., Herod, A. A., James, R. G., *Carcinogens–A Comprehensive Survey*, vol. 3:, *Polynuclear Aromatic Hydrocarbons*, Raven Press, New York (1978), pp. 35–45.

Determination of polycyclic aromatic hydrocarbons in atmospheric particulate matter by gas chromatography–mass spectrometry and high-pressure liquid chromatography, Thomas, R. S., Lao, R. C., Wang, D. T., Robinson, D., Sakuma, T., *Carcinogens–A Comprehensive Survey*, vol. 3: *Polynuclear Aromatic Hydrocarbons*, Raven Press, New York (1978), pp. 9–19.

BENZO[*b*]FLUORANTHENE $C_{20}H_{12}$ (252)

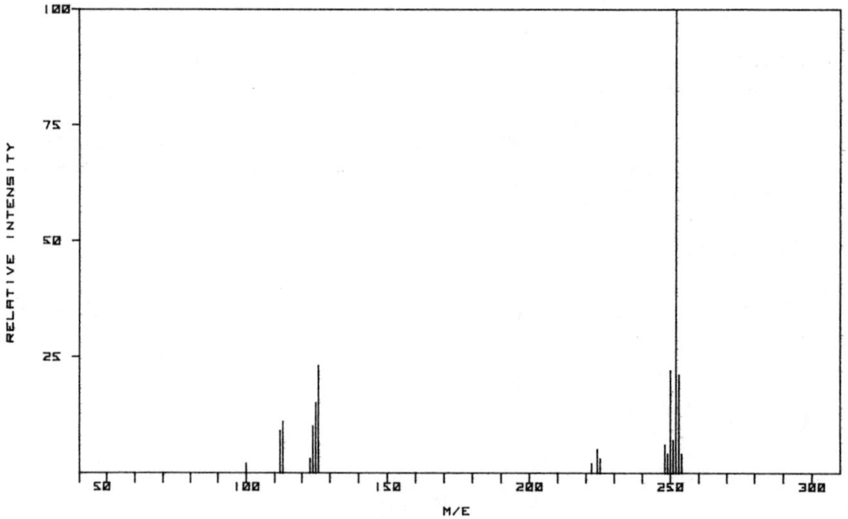

Spectral Data

Mass	Abundance	Mass	Abundance	Mass	Abundance
100	2.1	126	23.0	250	21.7
112	8.8	222	2.4	251	6.9
113	10.8	224	4.7	252	100.0
123	2.7	225	2.8	253	20.5
124	9.7	248	5.6	254	3.5
125	14.7	249	4.1		

Base–neutral extractable
CAS Name: Benz[*e*]acephenanthrylene
Synonyms
- 3,4-Benz[*e*]acephenanthrylene
- 3,4-Benzfluoranthrene
- Benzo[*e*]fluoranthene
- 2,3-Benzofluoranthene
- 3,4-Benzofluoranthene

CAS Registry No.: 205-99-2
ROTECS Ref.: CU14000

Major Ions: 252, 126, 250, 253, 125, 113, 124, 112
EPA Ions: 252, 253, 125

Selected Bibliography

Behavior of polycyclic aromatic hydrocarbons during their characterization by various indexes, Argirova, M., *Khig. Zdraveopaz.*, 18(3), 290–295 (1975).

Quantitative field desorption mass spectrometry. II. Mixtures of polycyclic hydrocarbons, Pfeifer, S., Beckey, H. D., Schulten, H. R., *Fresenius' Z. Anal. Chem.*, **284**(3), 193–195 (1977).

BENZO[*k*]FLUORANTHENE $C_{20}H_{12}$ (252)

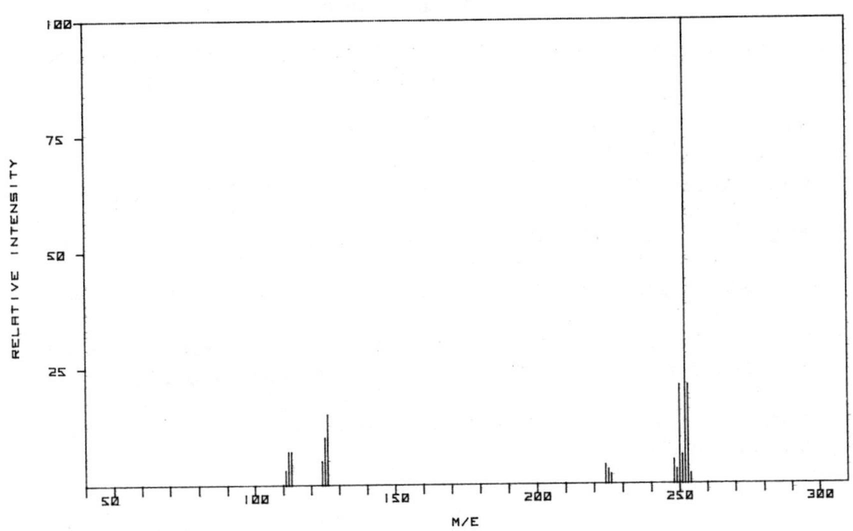

BENZO[*k*]FLUORANTHENE–*continued*

Spectral Data

Mass	Abundance	Mass	Abundance	Mass	Abundance
111	2.5	224	4.1	250	20.9
112	6.5	225	2.9	251	6.2
113	7.1	226	2.1	252	100.0
124	4.8	248	5.1	253	20.7
125	10.2	249	3.3	254	2.1
126	14.9				

Base–neutral extractable
CAS Name: Benzo[*k*]fluoranthene
Synonyms
- 8,9-Benzofluoranthene
- 11,12-Benzofluoranthene
- 11,12-Benzo[*k*]fluoranthene
- Dibenzo[*b,jk*]fluorene
- 2,3,1′,8′-Binaphthylene

CAS Registry No.: 207-08-9
ROTECS Ref.: DF63500

Major Ions: 252, 250, 253, 126, 125, 113, 112, 251
EPA Ions: 252, 253, 125

Selected Bibliography

Mass, metastable, and ion kinetic energy spectra of some polycyclic hydrocarbons found in environmental samples, Lao, R. C., Thomas, R. S., Monkman, J. L., Pottie, R. F., *Adv. Mass Spectrom.*, **6**, 129–136 (1974).

Behavior of polycyclic aromatic hydrocarbons during their characterization by various indexes, Argirova, M., *Khig. Zdraveopaz.*, **18**(3), 290–295 (1975).

Gas–liquid chromatographic evaluation and gas-chromatography/mass spectrometric application of new high-temperature liquid crystal stationary phases for polycyclic aromatic hydrocarbon separations, Janini, G. M., Muschik, G. M., Schroer, J. A., Zielinsk, W. L., Jr., *Anal. Chem.*, **48**(13), 1879–1883 (1976).

Characterization of sulfur-containing polycyclic aromatic compounds in carbon blacks, Lee, M. L., Hites, R. A., *Anal. Chem.*, **48**(13), 1890–1893 (1976).

Determination of polycyclic aromatic hydrocarbons in atmospheric particulate matter by gas chromatography–mass spectrometry and high-pressure liquid chromatography, Thomas, R. S., Lao, R. C., Wang, D. T., Robinson, D., Sakuma, T., *Carcinogens–A Comprehensive Survey*, vol. 3: *Polynuclear Aromatic Hydrocarbons*, Raven Press, New York (1978), pp. 9–19.

BENZO[*ghi*]PERYLENE $C_{22}H_{12}$ (276)

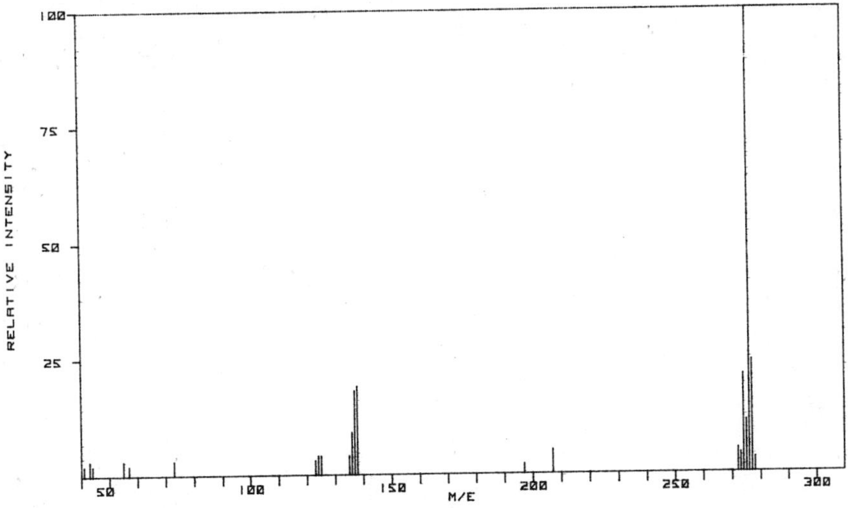

Spectral Data

Mass	Abundance	Mass	Abundance	Mass	Abundance
41	2.0	125	4.0	272	5.2
43	3.0	135	3.8	273	4.3
44	2.4	136	8.9	274	21.3
55	2.5	137	17.8	275	11.4
57	2.4	138	19.4	276	100.0
73	3.2	197	2.1	277	23.7
123	2.6	207	4.5	278	2.5
124	3.9				

Base–neutral extractable
CAS Name: Benzo[*ghi*]perylene
Synonym: 1,12-Benzoperylene

CAS Registry No.: 191-24-2

Major Ions: 276, 277, 274, 138, 137, 275, 136, 272
EPA Ions: 276, 138, 277

Selected Bibliography

Behavior of polycyclic aromatic hydrocarbons during their characterization by various indexes, Argirova, M., *Khig. Zdraveopaz.*, 18(3), 290–295 (1975).

Gas–liquid chromatographic evaluation and gas-chromatography/mass spectrometric application of new high-temperature liquid crystal stationary phases for polycyclic aromatic hydrocarbon separations, Janini, G. M., Muschik, G. M., Schroer, J. A., Zielinsk, W. L., Jr., *Anal. Chem.*, 48(13), 1879–1883 (1976).

Characterization of sulfur-containing polycyclic aromatic compounds in carbon blacks, Lee, M. L., Hites, R. A., *Anal. Chem.*, 48(13), 1890–1893 (1976).

BENZO[*a*]PYRENE $C_{20}H_{12}$ (252)

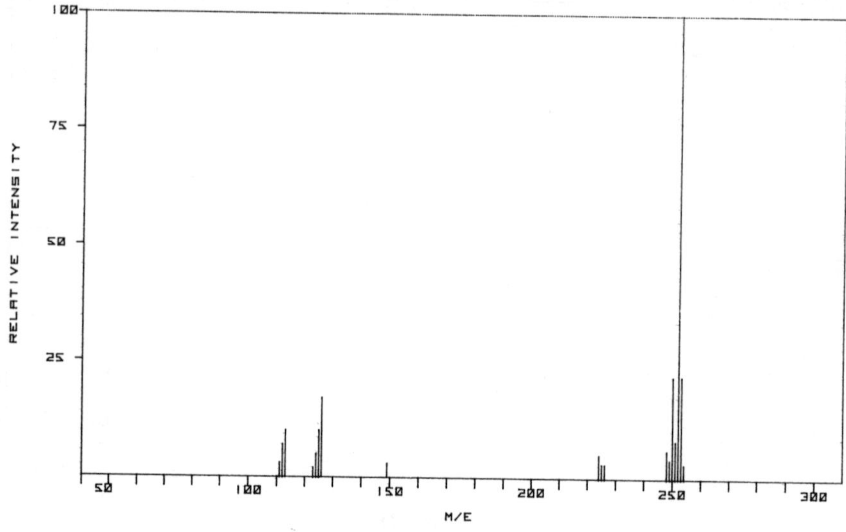

Spectral Data

Mass	Abundance	Mass	Abundance	Mass	Abundance
111	2.9	126	16.9	249	3.8
112	6.8	149	3.4	250	22.1
113	9.7	224	4.5	251	8.2
123	2.4	225	3.1	252	100.0
124	5.4	226	3.3	253	22.0
125	10.4	248	5.8	254	2.9

Base–neutral extractable
CAS Name: Benzo[*a*]pyrene
Synonyms

3,4-Benzopirene	Benzpyrene	BP
3,4-Benzopyrene	3,4-Benzpyrene	3,4-BP
6,7-Benzopyrene	3,4-Benzypyrene	B(a)P
3,4-Benzpyren	Benz[*a*]pyrene	

CAS Registry No.: 50-32-8
ROTECS Ref.: DJ36750
Merck Index Ref.: 1113

Major Ions: 252, 250, 253, 126, 125, 113, 251, 112
EPA Ions: 252, 253, 125

Selected Bibliography

Mass, metastable, and ion kinetic energy spectra of some polycyclic hydrocarbons found in environmental samples, Lao, R. C., Thomas, R. S., Monkman, J. L., Pottie, R. F., *Adv. Mass Spectrom.*, **6**, 129–136 (1974).

Gas–liquid chromatographic evaluation and gas-chromatography/mass spectrometric application of new high-temperature liquid crystal stationary phases for polycyclic aromatic hydrocarbon separations, Janini, G. M., Muschik, G. M., Schroer, J. A., Zielinsk, W. L., Jr., *Anal. Chem.*, **48**(13), 1879–1883 (1976).

A comparison of some chromatographic methods for estimation of polynuclear aromatic hydrocarbons in pollutants, Burchill. P., Herod, A. A., James, R. G., *Carcinogens–A Comprehensive Survey*, vol. 3: *Polynuclear Aromatic Hydrocarbons*, Raven Press, New York (1978), pp. 35–45.

Determination of polynuclear aromatic hydrocarbons contaminated with chlorinated hydrocarbon pesticides, Negishi, T., *Bull. Environ. Contam. Toxicol.*, **19**(5), 545–548 (1978).

Determination of polycyclic aromatic hydrocarbons in atmospheric particulate matter by gas chromatography-mass spectrometry and high-pressure liquid chromatography, Thomas, R. S., Lao, R. C., Wang, D. T., Robinson, D., Sakuma, T., *Carcinogens–A Comprehensive Survey*, vol. 3: *Polynuclear Aromatic Hydrocarbons*, Raven Press, New York (1978), pp. 9–19.

CHRYSENE $C_{18}H_{12}$ (228)

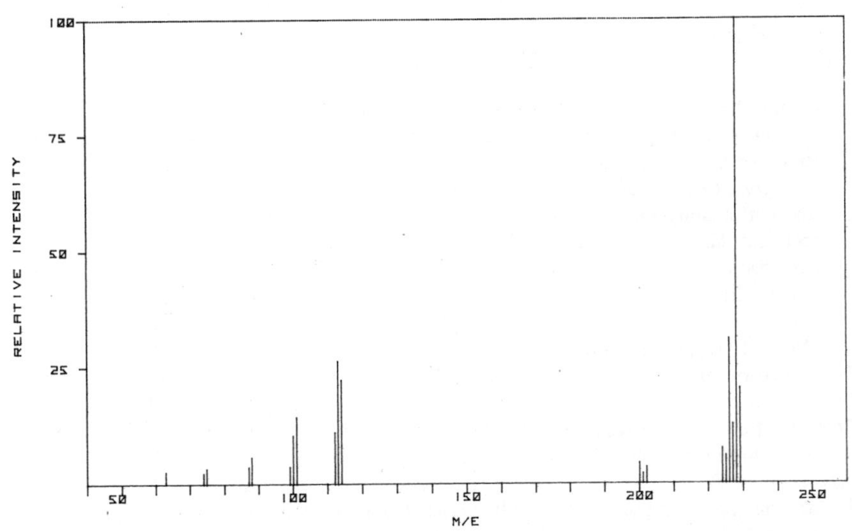

CHRYSENE–*continued*

Spectral Data

Mass	Abundance	Mass	Abundance	Mass	Abundance
63	2.6	101	14.3	224	7.4
74	2.2	112	11.0	225	5.8
75	3.3	113	26.3	226	30.9
87	3.6	114	22.2	227	12.5
88	5.7	200	4.3	228	100.0
99	3.6	201	2.1	229	20.3
100	10.3	202	3.4		

Base–neutral extractable
CAS Name: Chrysene
Synonyms
Benzo[*a*]phenanthrene
1,2-Benzophenanthrene
1,2-Benzphenanthrene
Benz[*a*]phenanthrene
1,2,5,6-Dibenzonaphthalene

CAS Registry No.: 218-01-9 (formerly 27274-05-1)
ROTECS Ref.: GC07000
Merck Index Ref.: 2252

Major Ions: 228, 226, 113, 114, 229, 227, 112, 100
EPA Ions: 228, 226, 229

Selected Bibliography

Gas–liquid chromatographic evaluation and gas-chromatography/mass spectrometric application of new high-temperature liquid crystal stationary phases for polycyclic aromatic hydrocarbon separations, Janini, G. M., Muschik, G. M., Schroer, J. A., Zielinsk, W. L., Jr., *Anal. Chem.*, 48(13), 1879–1883 (1976).

Gas chromatography–mass spectrometry of simulated arson residue using gasoline as an accelerant, Mach, M. H., *J. Forensic Sci.*, 22(2), 348–357 (1977).

A comparison of some chromatographic methods for estimation of polynuclear aromatic hydrocarbons in pollutants, Burchill, P., Herod, A. A., James, R. G., *Carcinogens–A Comprehensive Survey*, vol. 3: *Polynuclear Aromatic Hydrocarbons*, Raven Press, New York (1978), pp. 35–45.

Determination of polynuclear aromatic hydrocarbons contaminated with chlorinated hydrocarbon pesticides, Negishi, T., *Bull. Environ. Contam. Toxicol.*, 19(5), 545–548 (1978).

Determination of polycyclic aromatic hydrocarbons in atmospheric particulate matter by gas chromatography–mass spectrometry and high-pressure liquid chromatography, Thomas, R. S., Lao, R. C., Wang, D. T., Robinson, D., Sakuma, T., *Carcinogens–A Comprehensive Survey*, vol. 3: *Polynuclear Aromatic Hydrocarbons*, (1978), pp. 9–19.

DIBENZO[*a*, *h*]ANTHRACENE $C_{22}H_{14}$ (278)

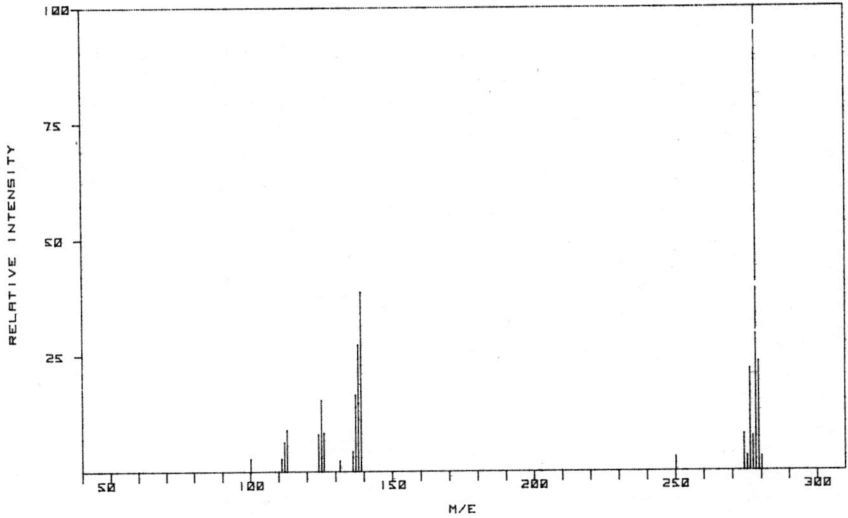

Spectral Data

Mass	Abundance	Mass	Abundance	Mass	Abundance
100	2.6	131	2.2	275	3.1
111	2.6	136	4.2	276	21.9
112	6.2	137	16.3	277	7.3
113	8.8	138	27.2	278	100.0
124	7.9	139	38.6	279	23.4
125	15.2	250	2.9	280	2.9
126	8.2	274	7.7		

Base–neutral extractable
CAS Name: Dibenz[*a*, *h*]anthracene
Synonyms

- 1,2:5,6-Benzanthrancene
- DBA
- DB(a,h)A
- Dibenzanthracene
- 1,2:5,6-Dibenz[*a*]anthracene
- Dibenzanthrene
- 1,2:5,6-Dibenzoanthracene

CAS Registry No.: 53-70-3
ROTECS Ref.: HN26250
Merck Index Ref.: 2971

Major Ions: 278, 139, 138, 279, 276, 137, 125, 113
EPA Ions: 278, 139, 279

DIBENZO[*a*,*h*]ANTHRACENE–*continued*

Selected Bibliography

Gas–liquid chromatographic evaluation and gas-chromatography/mass spectrometric application of new high-temperature liquid crystal stationary phases for polycyclic aromatic hydrocarbon separations, Janini, G. M., Muschik, G. M., Schroer, J. A., Zielinsk, W. L., Jr., *Anal. Chem.*, 48(13), 1879–1883 (1976).

Mixed charge exchange–chemical ionization mass spectrometry of polycyclic aromatic hydrocarbons, Lee, M. L., Hites, R. A., *J. Am. Chem. Soc.*, 99(6), 2008–2009 (1977).

Determination of polycyclic aromatic hydrocarbons in atmospheric particulate matter by gas chromatography–mass spectrometry and high-pressure liquid chromatography, Thomas, R. S., Lao, R. C., Wang, D. T., Robinson, D., Sakuma, T., *Carcinogens–A Comprehensive Survey*, vol. 3: *Polynuclear Aromatic Hydrocarbons*, Raven Press, New York (1978), pp. 9–19.

Determination of polynuclear aromatic hydrocarbons contaminated with chlorinated hydrocarbon pesticides, Negishi, T., *Bull. Environ. Contam. Toxicol.*, 19(5), 545–548 (1978).

FLUORENE $C_{13}H_{10}$ (166)

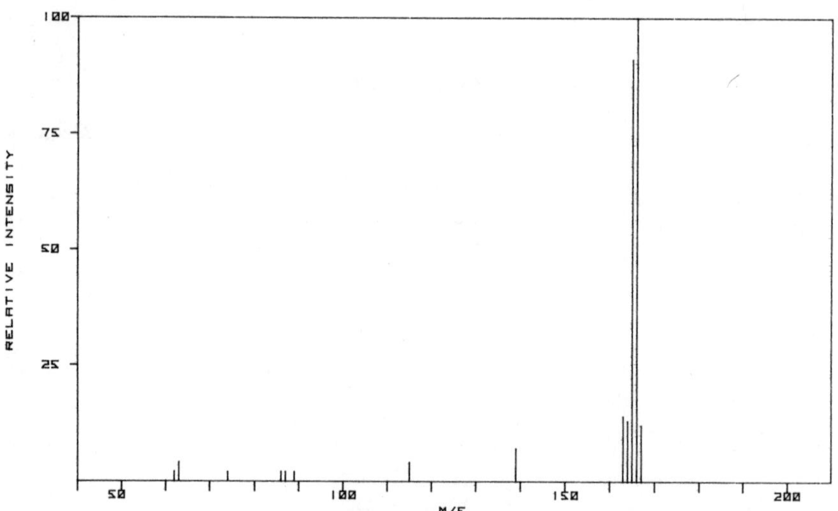

Spectral Data

Mass	Abundance	Mass	Abundance	Mass	Abundance
62	2.2	89	2.0	164	12.6
63	3.9	115	4.2	165	90.9
74	2.1	139	6.5	166	100.0
86	2.0	163	14.2	167	12.4
87	2.4				

Base–neutral extractable
CAS Name: Fluorene
Synonyms
o-Biphenylenemethane
Diphenylenemethane
2,2′-Methylenebiphenyl

CAS Registry No.: 86-73-7
Merck Index Ref.: 4037

Major Ions: 166, 165, 163, 164, 167, 139, 115, 63
EPA Ions: 166, 165, 167

Selected Bibliography

Data processing of low-energy ionization mass spectra and type determination of aromatic hydrocarbons, Nishishita, T., Yoshihara, M., Oshima, S., *Maruzen Sekiyu Giho*, (16), 99–104 (1971).

Comparative study of the mass spectra of stilbene and fluorene, Guesten, H., Klasinc, L., Marsel, J., Milivojevic, D., *Org. Mass Spectrom.*, **6**(2), 175–178 (1972).

Multiply charged ions in the mass spectra of aromatics, Engel, R., Halpern, D., Funk, B. A., *Org. Mass Spectrom.*, 7(2), 177–183 (1973).

Mass spectral studies of fluorenone and fluorene, Marshall, J. L., Kattner, R. M., *Tex. J. Sci.*, **30**(1), 77–84 (1978).

Determination of polycyclic aromatic hydrocarbons in atmospheric particulate matter by gas chromatography–mass spectrometry and high-pressure liquid chromatography, Thomas, R. S., Lao, R. C., Wang, D. T., Robinson, D., Sakuma, T., *Carcinogens–A Comprehensive Survey*, vol. 3: *Polynuclear Aromatic Hydrocarbons*, Raven Press, New York (1978), pp. 9–19.

INDENO[1,2,3-*cd*]PYRENE $C_{22}H_{12}$ (276)

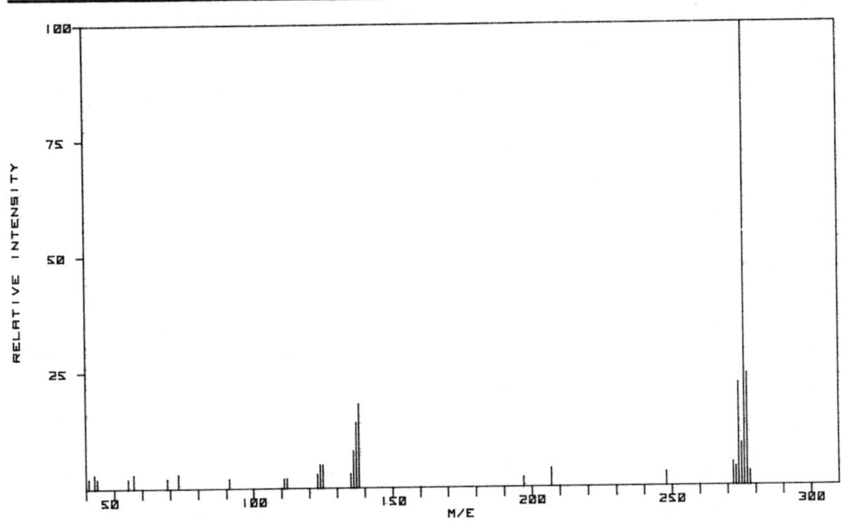

INDENO[1,2,3-*cd*]PYRENE–*continued*

Spectral Data

Mass	Abundance	Mass	Abundance	Mass	Abundance
41	2.3	112	2.3	207	3.7
43	2.5	123	2.9	248	2.6
44	2.0	124	4.8	272	5.1
55	2.3	125	5.2	273	4.1
57	2.5	135	3.2	274	21.5
69	2.4	136	7.8	275	9.2
73	3.1	137	13.9	276	100.0
91	2.1	138	18.2	277	24.1
111	2.3	197	2.2	278	3.3

Base–neutral extractable
CAS Name: Indeno[1,2,3-*cd*]pyrene
Synonyms

- IP
- 2,3-Phenylenepyrene
- 1,10-(1,2-Phenylene)pyrene
- *o*-Phenylenepyrene
- 2,3-*o*-Phenylenepyrene
- 1,10-(*o*-Phenylene)pyrene

CAS Registry No.: 193-39-5
ROTECS Ref.: UR26250

Major Ions: 276, 277, 274, 138, 137, 275, 136, 125
EPA Ions: 276, 138, 277

Selected Bibliography

Characterization of sulfur-containing polycyclic aromatic compounds in carbon blacks, Lee, M. L., Hites, R. A., *Anal. Chem.*, 48(13), 1890–1893 (1976).

Quantitative field desorption mass spectrometry. II. Mixtures of polycyclic hydrocarbons, Pfeifer, S., Beckey, H. D., Schulten, H. R., *Fresenius' Z. Anal. Chem.*, 284(3), 193–195 (1977).

PHENANTHRENE $C_{14}H_{10}$ (178)

Spectral Data

Mass	Abundance	Mass	Abundance	Mass	Abundance
63	4.0	88	7.3	175	2.2
74	2.5	89	8.8	176	17.9
75	4.9	150	5.1	177	10.1
76	10.3	151	6.8	178	100.0
87	2.5	152	9.4	179	14.2

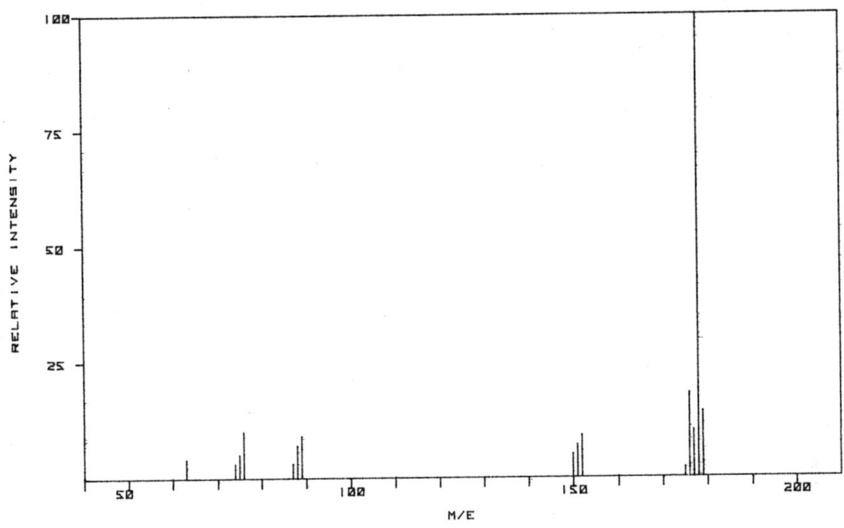

Base–neutral extractable
CAS Name: Phenanthrene

CAS Registry No.: 85-01-8
ROTECS Ref.: SF71750
Merck Index Ref.: 6996

Major Ions: 178, 176, 179, 76, 177, 152, 89, 88
EPA Ions: 178, 179, 176

Selected Bibliography

Multiply charged ions in the mass spectra of aromatics, Engel, R., Halpern, D., Funk, B. A., *Org. Mass Spectrom.*, 7(2), 177–183 (1973).

Analysis of complex polycyclic aromatic hydrocarbon mixtures by computerized GC/MS, Hites, R. A., *Prepr., Div. Pet. Chem., Am. Chem. Soc.*, 20(4), 824–882 (1975).

Selection of a method for determining the composition of aromatic hydrocarbons in high-boiling petroleum fractions, Siryuk, A. G., Barabadze, Sh. Sh., *Khim Tekhnol. Topl. Masel*, (10), 54–56 (1977).

Determination of polynuclear aromatic hydrocarbons contaminated with chlorinated hydrocarbon pesticides, Negishi, T., *Bull. Environ. Contam. Toxicol.*, 19(5), 545–548 (1978).

Determination of polycyclic aromatic hydrocarbons in atmospheric particulate matter by gas chromatography–mass spectrometry and high-pressure liquid chromatography, Thomas, R. S., Lao, R. C., Wang, D. T., Robinson, D., Sakuma, T., *Carcinogens–A Comprehensive Survey*, vol. 3: *Polynuclear Aromatic Hydrocarbons*, Raven Press, New York (1978), pp. 9–19.

PYRENE $C_{16}H_{10}$ (202)

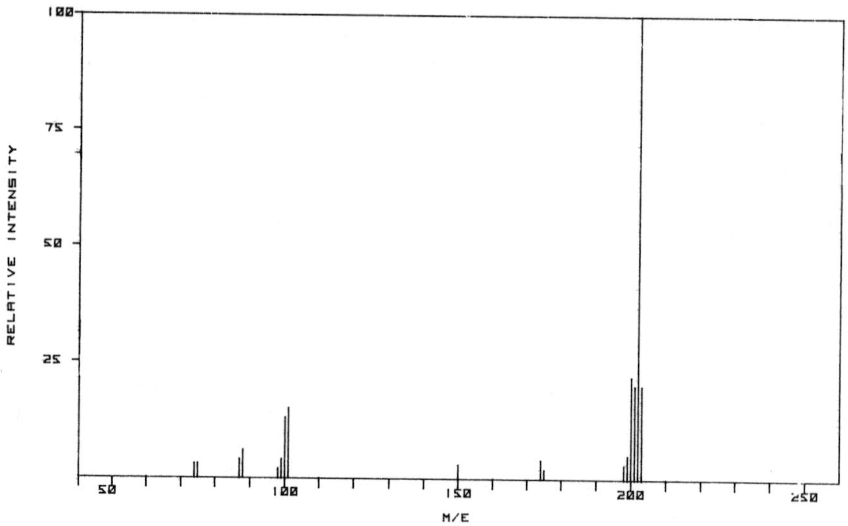

Spectral Data

Mass	Abundance	Mass	Abundance	Mass	Abundance
74	2.6	100	13.0	199	4.5
75	2.6	101	15.4	200	22.2
87	4.3	150	2.6	201	19.7
88	5.5	174	3.7	202	100.0
98	2.0	175	2.0	203	19.5
99	3.5	198	2.8		

Base–neutral extractable
CAS Name: Pyrene
Synonyms
    Benzo[*def*]phenanthrene
    β-Pyrene

CAS Registry No.: 129-00-0 (formerly 4820-01-3)
ROTECS Ref.: UR24500
Merck Index Ref.: 7746

Major Ions: 202, 200, 201, 203, 101, 100, 88, 199
EPA Ions: 202, 101, 100

Selected Bibliography

Gas–liquid chromatographic evaluation and gas-chromatography/mass spectrometric application of new high-temperature liquid crystal stationary phases for polycyclic aromatic hydrocarbon separations, Janini, G. M., Muschik, G. M., Schroer, J. A., Zielinsk, W. L., Jr., *Anal. Chem.*, 48(13), 1879–1883 (1976).

Gas chromatography–mass spectrometry of simulated arson residue using gasoline as an accelerant, Mach, M. H., *J. Forensic Sci.*, 22(2), 348–357 (1977).

Determination of polynuclear aromatic hydrocarbons contaminated with chlorinated hydrocarbon pesticides, Negishi, T., *Bull. Environ. Contam. Toxicol.*, 19(5), 545–548 (1978).

2,3,7,8-TETRACHLORODIBENZO-*p*-DIOXIN

This teratogen is so toxic that the U.S. Environmental Protection Agency recommends that analytical laboratories do not acquire the authentic material for use as a standard. It is present as an impurity in chlorinated phenols and related compounds. It is a particular hazard in waste materials produced as byproducts in the manufacture of such compounds.

2,3,7,8-TETRACHLORODIBENZO-*p*-DIOXIN $C_{12}H_4Cl_4O_2$ (320)

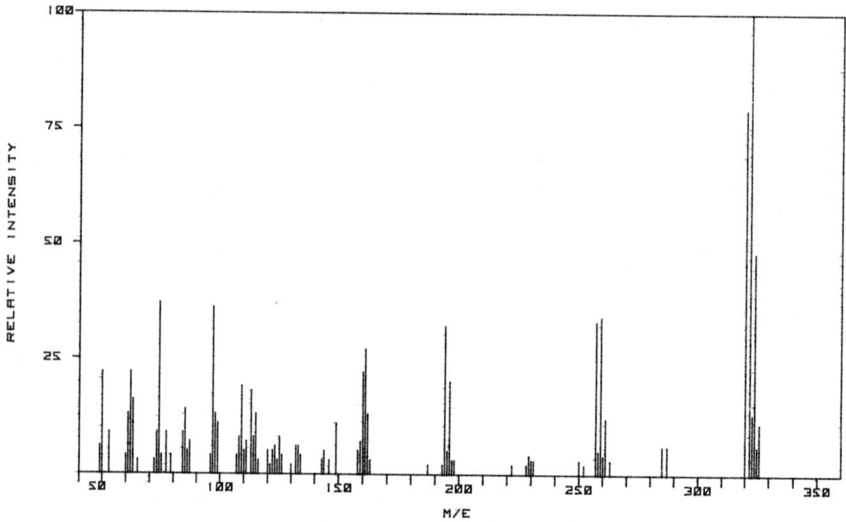

Spectral Data

Mass	Abundance	Mass	Abundance	Mass	Abundance
49	5.6	74	37.4	98	13.0
50	22.3	75	3.7	99	10.9
53	8.5	77	8.7	107	3.7
60	3.9	79	4.1	108	7.6
61	13.4	84	9.1	109	19.0
62	21.5	85	13.8	110	4.7
63	16.1	86	4.5	111	7.4
65	3.1	87	7.4	113	18.0
72	3.3	96	4.1	114	7.9
73	9.3	97	36.0	115	12.6

Spectral Data—*continued*

Mass	Abundance	Mass	Abundance	Mass	Abundance
116	3.1	159	7.0	250	3.1
120	5.0	160	22.1	252	2.3
121	2.3	161	26.7	257	32.9
122	5.4	162	13.4	258	4.8
123	6.2	163	3.1	259	34.3
124	3.1	187	2.1	260	4.3
125	7.6	193	2.3	261	11.8
126	4.3	194	31.6	263	2.7
130	2.3	195	5.0	285	6.0
132	5.6	196	19.6	287	5.8
133	6.0	197	3.1	320	78.5
134	3.9	198	3.1	322	100.0
143	2.7	222	2.3	323	13.4
144	5.4	228	2.1	324	47.9
146	2.9	229	3.5	325	6.4
149	11.2	230	2.5	326	10.5
158	5.0	231	3.3		

Base–neutral extractable

CAS Name: Dibenzo-*p*-dioxin, 2,3,7,8-tetrachloro-

Synonyms

Dioxin	TCDD
NCI-C03714	2,3,7,8-Tetrachlorodibenzo-1,4-dioxin
TCDBD	

CAS Registry No.: 1746-01-6 (formerly 56795-67-6)
ROTECS Ref.: HP35000

Major Ions: 322, 320, 324, 74, 97, 259, 257, 194
EPA Ions: 322, 320

Selected Bibliography

Fragmentation of dibenzo-*p*-dioxin and its derivatives under electron impact, Buu-Hoi, N. P., Saint-Ruf, G., Mangane, M., *J. Heterocycl. Chem.*, 9(3), 691–693 (1972).

Mass spectrometry in toxicology. 1. Calculation of theoretical isotopic abundances and optimum ion dwell time for a 2,3,7,8-tetrachlorodibenzo-*p*-dioxin pharmacokinetic study, Reynolds, W. D., Delongchamp, R., *Biomed. Mass Spectrom.*, 2(5), 276–278 (1975).

Identification of substitution patterns in polychlorinated dibenzo-*p*-dioxins (PCDDs) by mass spectrometry, Buser, H. R., Rappe, C., *Chemosphere*, 7(2), 199–211 (1978).

Determination of polychlorinated dibenzo-*p*-dioxins in biological samples by negative chemical ionization mass spectrometry, Hass, J. R., Friesen, M. D., Harvan, D. J., Parker, C. E., *Anal. Chem.*, **50**(11), 1474–1479 (1978).

Chlorinated benzyl phenyl ethers: a possible interference in the determination of chlorinated dibenzo-*p*-dioxins in 2,4,5-trichlorophenol and its derivatives, Shadoff, L. A., Blaser, W. W., Kocher, C. W., Fravel, H. G., *Anal. Chem.*, **50**(11), 1586–1588 (1978).

TETRACHLOROETHYLENE

1,1,2,2-Tetrachlorethene is employed as a solvent, and in dry cleaning and degreasing of metals. It is also used as an anthelmintic for hookworms, pinworms, ascarids, and tapeworms. It has been employed as a fumigant for insects and rodents.

1,1,2,2-TETRACHLOROETHENE C_2Cl_4 (164)

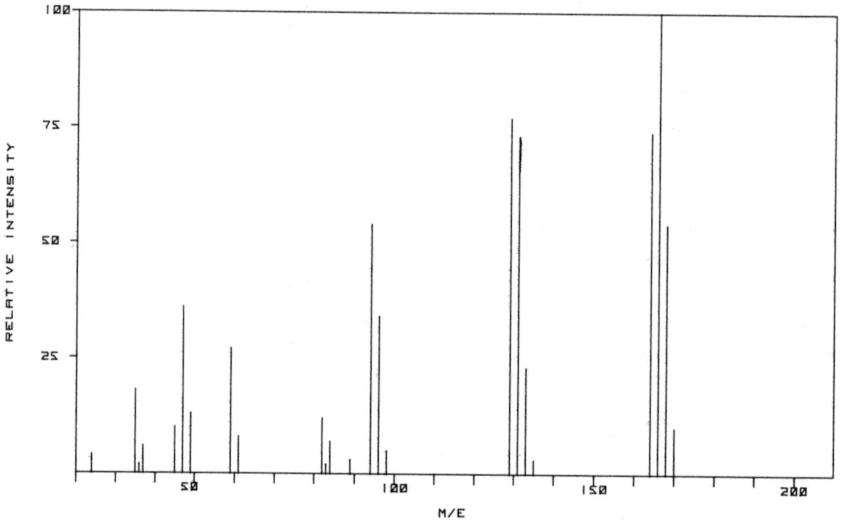

Spectral Data

Mass	Abundance	Mass	Abundance	Mass	Abundance
24	3.5	61	7.8	129	77.1
35	17.7	82	12.1	131	73.4
36	2.2	83	2.2	133	23.0
37	6.1	84	7.3	135	2.7
45	10.1	89	2.8	164	74.2
47	35.7	94	53.8	166	100.0
49	12.8	96	33.6	168	54.3
59	26.9	98	5.3	170	10.3

TETRACHLOROETHYLENE

Volatile
CAS Name: Ethylene, tetrachloro-
Synonyms

Ankilostin	NCI-C04580	Tetracap
Antisal 1	Nema	Tetraguer
Carbon bichloride	Perawin	Tetrachlorethylene
Carbon dichloride	Perc	Tetrachloroethene
Didakene	Perchloroethylene	Tetrachloroethylene
Ent 1,860	Perclene	Tetraleno
Ethylene tetrachloride	Persec	Tetralex
Fedal-Un	Tetlen	Tetropil

CAS Registry No.: 127-18-4
ROTECS Ref.: KX38500
Merck Index Ref.: 8907

Major Ions: 166, 129, 164, 131, 168, 94, 47, 96
EPA Ions: 129, 131, 164, 166

Selected Bibliography

Positive-ion mass spectra of some perhalogenated compounds, Contineanu, M. A., Grubel, K., *An. Univ. Bucuresti. Chim.*, **20**(2), 175–181 (1971).

Direct aqueous injection gas chromatography–mass spectrometry for analysis of organohalides in water at concentrations below the parts per billion level, Fujii, T., *J. Chromatogr.*, **139**(2), 297–302 (1977).

The determination of volatile organic compounds in city air by gas chromatography combined with standard addition, selective subtraction, infrared spectrometry and mass spectrometry, Louw, C. W., Richards, J. F., Faure, P. K., *Atmos. Environ.*, **11**(8), 703–717 (1977).

A field portable mass spectrometer for monitoring organic vapors, Meier, R. W., *J. Am. Ind. Hyg. Assoc.*, **39**(3), 233–239 (1978).

TOLUENE

Toluene is used as a solvent and in the manufacture of many organic compounds. In the laboratory, it is used as substitute for benzene, but it is a narcotic in high concentrations and may cause mild macrocytic anemia.

TOLUENE C_7H_8 (92)

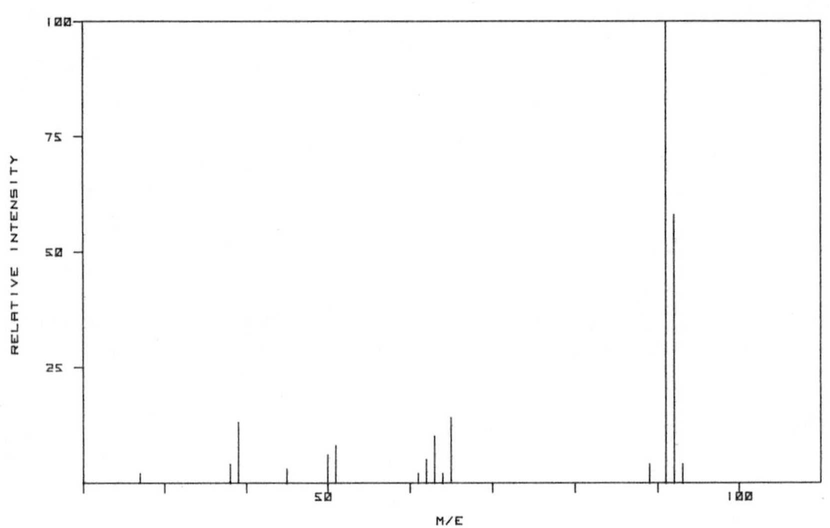

Spectral Data

Mass	Abundance	Mass	Abundance	Mass	Abundance
27	2.2	51	8.3	65	14.4
38	3.5	61	2.2	89	4.2
39	12.8	62	5.0	91	100.0
45	2.6	63	10.4	92	57.8
50	5.9	64	2.1	93	3.9

Volatile
CAS Name: Toluene
Synonyms

- Antisal 1a
- Methacide
- Methylbenzene
- Methylbenzol
- Phenylmethane
- Toluol

TOLUENE

CAS Registry No.: 108-88-3
ROTECS Ref.: XS52500
Merck Index Ref.: 9225

Major Ions: 91, 92, 65, 39, 63, 51, 50, 62
EPA Ions: 91, 92

Selected Bibliography

Doubly charged ions from labeled toluenes. Isotope effects, preference factors, and hydrogen randomization, Ast. T., Beynon, J. H., Cooks, R. G., *J. Am. Chem. Soc.*, **94**(6), 1834–1836 (1972).

Doubly-charged ion mass spectra of hydrocarbons, Ast. T., Beynon, J. H., Cooks, R. G., *Org. Mass Spectrom.*, **6**(7), 749–763 (1972).

Multiply charged ions in the mass spectra of aromatics, Engel, R., Halpern, D., Funk, B. A., *Org. Mass Spectrom.*, **7**(2), 177–183 (1973).

Analysis of leek volatiles by headspace condensation, Schreyen, L., Dirinck, R., Van Wassenhove, F., Schamp, N., *J. Agric. Food Chem.*, **24**(6), 1147–1152 (1976).

TOXAPHENE

Toxaphene is a mixture of compounds formed by chlorination of camphene (67–69% chlorine). It is a powerful contact and stomach insecticide with some acaricidal action. It is not phytotoxic, is toxic to fish, and is moderately toxic to mammals. It is used against army worms, boll weevil, bollworm, cotton aphid, cotton fleahopper, cotton leafworm, grasshopper, rapid plant bug, southern green stink bug, tarnished plant bug, and thrips.

TOXAPHENE **(mixture)**

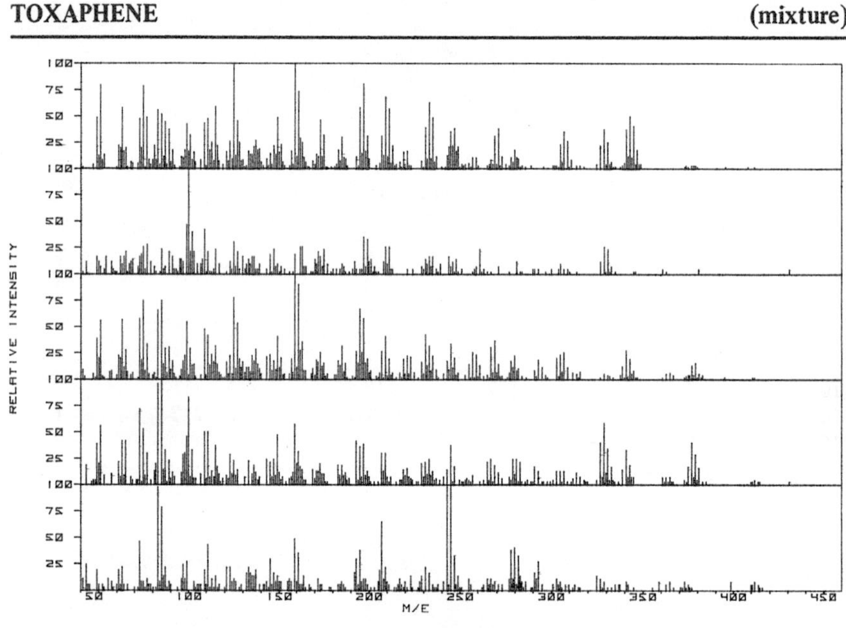

Spectral Data

Peak 1

Mass	Abundance	Mass	Abundance	Mass	Abundance
41	2.1	53	13.4	65	19.6
47	4.1	60	2.1	66	2.1
49	48.5	61	21.6	68	6.2
50	12.4	62	19.6	69	5.2
51	79.4	63	57.7	72	5.2
52	8.2	64	16.5	73	47.4

Peak 1–*continued*

Mass	Abundance	Mass	Abundance	Mass	Abundance
74	19.6	128	24.7	181	2.1
75	78.4	129	7.2	182	4.1
76	2.1	130	8.2	183	17.5
77	48.5	131	2.1	184	7.2
78	8.2	132	3.1	185	29.9
79	4.1	133	16.5	186	10.3
80	8.2	134	13.4	187	9.3
81	21.6	135	13.4	188	3.1
82	8.2	136	20.6	193	11.3
83	55.7	137	26.8	194	3.1
84	4.1	138	17.5	195	57.7
85	51.5	139	18.6	196	14.4
86	11.3	140	6.2	197	80.4
87	44.3	141	3.1	198	16.5
88	4.1	143	9.3	199	30.9
89	37.1	144	3.1	200	9.3
90	6.2	145	14.4	204	3.1
91	17.5	146	4.1	206	5.2
92	3.1	147	21.6	207	30.9
95	2.1	148	13.4	208	12.4
96	11.3	149	48.5	209	68.0
97	10.3	150	15.5	210	11.3
98	17.5	151	22.7	211	56.7
99	42.3	152	6.2	212	5.2
100	16.5	153	3.1	213	21.6
101	32.0	156	3.1	214	3.1
102	8.2	157	16.5	215	4.1
103	15.5	158	4.1	217	6.2
104	6.2	159	100.0	218	2.1
107	8.2	160	11.3	219	15.5
109	43.3	161	73.2	220	4.1
110	3.1	162	28.9	221	16.5
111	47.4	163	24.7	222	3.1
112	17.5	164	10.3	223	3.1
113	24.7	165	6.2	229	7.2
114	7.2	166	2.1	230	2.1
115	58.8	167	2.1	231	39.2
116	21.6	169	7.2	232	9.3
117	7.2	170	4.1	233	62.9
119	4.1	171	13.4	234	9.3
121	15.5	172	11.3	235	48.5
122	7.2	173	46.4	236	6.2
123	25.8	174	11.3	237	11.3
124	9.3	175	32.0	240	2.1
125	100.0	176	2.1	241	2.1
126	12.4	177	2.1	243	12.4
127	45.4	179	2.1	244	20.6

TOXAPHENE–*continued*

Peak 1–*continued*

Mass	Abundance	Mass	Abundance	Mass	Abundance
245	35.1	280	17.5	332	4.1
246	21.6	281	10.3	333	6.2
247	38.1	282	9.3	334	2.1
248	16.5	283	2.1	335	2.1
249	20.6	284	3.1	338	2.1
250	2.1	286	2.1	339	3.1
251	4.1	289	3.1	340	5.2
252	3.1	296	2.1	341	37.1
253	4.1	301	3.1	342	11.3
255	5.2	302	3.1	343	49.5
256	2.1	303	3.1	344	10.3
257	10.3	304	2.1	345	40.2
259	10.3	305	22.7	346	4.1
260	3.1	306	6.2	347	17.5
265	3.1	307	35.1	348	4.1
266	4.1	309	25.8	349	4.1
267	8.2	310	2.0	373	2.1
268	4.1	311	8.2	374	3.1
269	20.9	314	3.1	375	2.1
270	4.1	316	3.1	377	3.1
271	38.1	318	2.1	378	3.1
272	3.1	325	4.1	379	3.1
273	11.3	327	21.6	380	2.1
275	4.1	329	37.1	395	2.1
277	3.1	330	4.1	408	2.1
278	10.3	331	24.7	411	2.1
279	6.2				

Peak 2

Mass	Abundance	Mass	Abundance	Mass	Abundance
41	2.3	66	2.3	86	4.7
43	11.6	67	9.3	87	7.0
49	16.3	68	11.6	89	20.9
50	11.6	69	14.0	91	16.3
51	7.0	72	7.0	92	4.7
53	4.7	73	16.3	93	4.7
54	16.3	74	18.6	94	2.3
57	11.6	75	25.6	95	14.0
58	4.7	76	4.7	96	14.0
59	2.3	77	27.9	97	11.6
60	2.3	79	11.6	99	46.5
62	16.3	81	7.0	100	100.0
63	9.3	82	2.3	101	20.9
64	16.3	84	2.3	102	39.5
65	20.9	85	23.3	103	20.9

Peak 2–*continued*

Mass	Abundance	Mass	Abundance	Mass	Abundance
105	9.3	170	14.0	239	9.3
107	9.3	171	9.3	244	16.3
108	14.0	172	20.9	245	7.0
109	41.9	173	16.3	246	11.6
110	2.3	174	7.0	247	2.3
111	20.9	175	23.3	248	14.0
112	2.3	177	9.3	249	2.3
113	2.3	179	2.3	250	4.7
114	4.7	180	4.7	257	2.3
115	23.3	182	4.7	258	4.7
117	9.3	184	4.7	259	7.0
121	2.3	185	9.3	261	23.3
123	11.6	186	7.0	262	2.3
124	4.7	187	2.3	263	4.7
125	30.2	189	2.3	265	2.3
126	7.0	191	4.7	267	2.3
127	20.9	192	11.6	271	7.0
128	4.7	193	4.7	277	2.3
130	16.3	195	4.7	281	11.6
131	4.7	196	4.7	283	2.3
132	9.3	197	34.9	284	2.3
133	14.0	198	7.0	290	4.7
134	9.3	199	32.6	291	4.7
135	16.3	200	11.6	293	4.7
136	16.3	201	14.0	297	2.3
137	4.7	202	2.3	300	4.7
138	4.7	203	7.0	304	4.7
139	9.3	204	2.3	305	9.3
143	4.7	205	2.3	307	4.7
145	18.6	208	11.6	309	4.7
146	2.3	209	25.6	310	2.3
147	23.3	210	9.3	314	2.3
148	7.0	211	25.6	327	11.6
149	9.3	212	4.7	329	25.6
150	2.3	213	4.7	330	2.3
151	7.0	221	4.7	331	23.3
153	4.7	224	4.7	332	2.3
156	2.3	229	7.0	333	7.0
158	2.3	230	4.7	335	2.3
159	18.6	231	14.0	345	2.3
162	25.6	232	9.3	346	2.3
163	25.6	233	16.3	361	4.7
164	7.0	234	7.0	363	2.3
165	7.0	235	16.3	381	4.7
168	2.3	237	4.7	431	4.7
169	2.3	238	4.7		

TOXAPHENE–*continued*

Peak 3

Mass	Abundance	Mass	Abundance	Mass	Abundance
41	9.1	100	4.0	152	4.0
42	3.0	101	29.3	153	5.1
47	5.1	102	13.1	156	5.1
48	2.0	103	14.1	157	19.2
49	38.4	104	3.0	158	6.1
50	20.2	107	4.0	159	100.0
51	55.6	109	47.5	160	11.1
52	3.0	110	4.0	161	89.9
53	2.0	111	41.4	162	27.3
56	7.1	112	13.1	163	35.4
57	8.1	113	23.2	164	8.1
60	4.0	114	16.2	165	8.1
61	22.2	115	31.3	168	5.1
62	20.2	116	14.1	169	10.1
63	56.6	117	6.1	170	4.0
64	11.1	118	4.0	171	18.2
65	27.3	119	2.0	172	16.2
66	7.1	120	2.0	173	25.3
68	4.0	122	5.1	174	12.1
69	6.1	123	22.2	175	15.2
72	2.0	124	5.1	176	3.0
73	57.6	125	77.8	177	5.1
74	18.2	126	10.1	181	4.0
75	74.7	127	53.5	182	4.0
76	8.1	128	19.2	183	17.2
77	33.3	129	11.1	184	13.1
78	5.1	130	16.2	185	31.3
80	4.0	131	6.1	186	7.1
81	6.1	132	11.1	187	15.2
82	2.0	133	11.1	188	2.0
83	65.7	134	7.1	191	3.0
85	74.7	135	22.2	192	2.0
86	14.1	136	17.2	193	26.3
87	29.3	137	28.3	194	15.2
88	6.1	138	14.1	195	66.7
89	30.3	139	10.1	196	25.3
90	9.1	140	5.1	197	57.6
91	17.2	143	21.2	198	15.2
92	5.1	145	23.2	199	19.2
93	2.0	146	8.1	200	6.1
95	3.0	147	17.2	201	8.1
96	9.1	148	9.1	204	3.0
97	17.2	149	40.4	205	2.0
98	22.2	150	11.1	206	6.1
99	54.5	151	20.2	207	26.3

Peak 3–*continued*

Mass	Abundance	Mass	Abundance	Mass	Abundance
208	2.0	258	2.0	314	5.1
209	40.4	259	23.2	315	2.0
210	3.0	260	3.0	316	7.1
211	19.2	261	13.1	327	2.0
212	4.0	263	3.0	329	5.1
213	6.1	265	4.0	331	4.0
215	2.0	266	2.0	332	3.0
217	6.1	267	30.3	334	2.0
218	2.0	268	6.1	338	4.0
220	5.1	269	36.4	339	12.1
221	22.2	270	6.1	340	2.0
222	5.1	271	14.1	341	27.3
223	21.2	272	3.0	342	3.0
225	6.1	273	3.0	343	19.2
227	2.0	276	2.0	344	5.1
229	16.2	277	6.1	345	8.1
230	5.1	278	17.2	346	2.0
231	42.4	279	11.1	347	2.0
232	13.1	280	22.2	363	5.1
233	31.1	281	8.1	365	6.1
234	8.1	282	10.1	367	4.0
235	22.2	284	3.0	373	2.0
237	9.1	286	2.0	374	3.0
238	2.0	291	8.1	375	4.0
241	5.1	292	3.0	376	4.0
243	17.2	293	18.2	377	13.1
244	9.1	295	11.1	378	4.0
245	33.3	297	4.0	379	15.2
246	11.1	298	2.0	380	2.0
247	20.2	303	20.2	381	5.1
248	4.0	304	3.0	382	2.0
249	6.1	305	23.2	383	4.0
253	4.0	306	7.1	395	2.0
255	14.1	307	25.3	411	2.0
256	2.0	309	11.1	412	2.0
257	25.3	312	6.1		

Peak 4

Mass	Abundance	Mass	Abundance	Mass	Abundance
43	18.6	53	8.6	68	7.1
46	2.9	57	10.0	69	4.3
47	4.3	61	21.4	71	4.3
48	4.3	62	7.1	72	4.3
49	38.6	63	41.4	73	71.4
50	20.0	64	8.6	74	2.9
51	55.7	65	41.4	75	52.9

TOXAPHENE–*continued*

Peak 4–***continued***

Mass	Abundance	Mass	Abundance	Mass	Abundance
76	8.6	136	18.6	198	8.6
77	30.0	137	11.4	199	12.9
79	4.3	138	4.3	200	7.1
81	12.9	139	7.1	201	2.9
82	20.0	143	24.3	203	2.9
83	95.7	145	21.4	205	2.9
85	100.0	146	7.1	207	30.0
86	14.3	147	24.3	208	12.9
87	32.9	148	8.6	209	30.0
88	2.9	149	47.1	210	7.1
89	22.9	150	11.4	211	2.9
90	5.7	151	4.3	212	7.1
91	11.4	152	7.1	215	2.9
92	7.1	154	2.9	217	4.3
96	11.4	156	5.7	218	4.3
97	28.6	157	7.1	219	14.3
98	30.0	158	11.4	220	8.6
99	45.7	159	57.1	221	15.7
100	82.9	160	20.0	222	7.1
101	7.1	161	31.4	223	5.7
102	32.9	162	17.1	224	4.3
103	5.7	163	14.3	226	2.9
104	5.7	164	4.3	227	2.9
107	10.0	165	4.3	229	20.0
109	50.0	169	14.3	230	7.1
110	5.7	170	5.7	231	21.4
111	50.0	171	12.9	232	10.0
112	7.1	172	12.9	233	24.3
113	12.9	173	20.0	234	8.6
114	7.1	174	10.0	235	12.9
115	37.1	175	10.0	236	4.3
116	17.1	176	2.9	237	5.7
117	10.0	177	2.9	239	4.3
118	2.9	183	18.6	241	7.1
119	5.7	184	8.6	243	14.3
121	8.6	185	18.6	245	37.1
122	5.7	186	8.6	246	2.9
123	28.6	187	11.4	247	17.1
124	10.0	188	4.3	250	4.3
125	22.9	189	5.7	251	2.9
126	8.6	193	41.4	252	2.9
127	14.3	194	4.3	253	4.3
131	7.1	195	35.7	254	5.7
133	22.9	196	7.1	255	4.3
135	10.0	197	38.6	256	2.9

Peak 4–*continued*

Mass	Abundance	Mass	Abundance	Mass	Abundance
257	8.6	300	2.9	342	7.1
261	2.9	301	4.3	343	18.6
263	4.3	303	12.9	344	7.1
265	21.4	305	12.9	345	7.1
266	4.3	306	2.9	361	7.1
267	24.3	307	12.9	362	2.9
268	2.9	309	2.9	363	7.1
269	18.6	311	2.9	364	2.9
271	11.4	312	5.7	365	7.1
273	5.7	314	11.4	366	2.9
277	7.1	316	7.1	367	2.9
278	4.3	318	4.3	373	4.3
279	24.3	319	2.9	374	2.9
280	10.0	320	2.9	375	17.1
281	24.3	325	4.3	377	40.0
282	2.9	327	40.0	378	7.1
283	21.4	328	7.1	379	28.6
284	2.9	329	58.6	380	4.3
286	2.9	330	11.4	381	15.7
287	4.3	331	34.3	383	2.9
289	2.9	332	4.3	385	2.9
290	2.9	333	17.1	410	2.9
291	17.1	334	4.3	411	2.9
293	12.9	335	2.9	412	4.3
295	2.9	337	2.9	414	2.9
296	2.9	339	14.3	415	2.9
298	2.9	341	32.9	430	2.9

Peak 5

Mass	Abundance	Mass	Abundance	Mass	Abundance
41	10.8	70	2.7	89	8.1
43	24.3	72	2.7	90	5.4
44	5.4	73	45.9	94	2.7
45	5.4	74	8.1	96	10.8
49	18.9	75	8.1	97	24.3
50	5.4	76	2.7	98	10.8
52	5.4	77	10.8	99	27.0
53	2.7	78	5.4	103	8.1
55	10.8	79	5.4	104	2.7
57	8.1	80	2.7	105	5.4
58	2.7	82	2.7	106	5.4
60	5.4	83	97.3	107	5.4
61	18.9	84	8.1	109	18.9
62	5.4	85	78.4	110	5.4
63	21.6	86	13.5	111	43.2
64	5.4	87	21.6	112	2.7
66	5.4	88	2.7	113	10.8

TOXAPHENE–*continued*

Peak 5–*continued*

Mass	Abundance	Mass	Abundance	Mass	Abundance
114	2.7	178	2.7	244	5.4
115	13.5	179	2.7	245	100.0
116	2.7	183	5.4	247	32.4
117	5.4	184	5.4	248	5.4
120	2.7	185	8.1	249	13.5
121	21.6	186	5.4	250	2.7
124	8.1	187	2.7	251	2.7
125	10.8	192	16.2	255	10.8
126	8.1	193	29.7	256	5.4
128	2.7	194	5.4	262	2.7
129	5.4	195	37.8	263	5.4
132	16.2	196	8.1	265	10.8
133	21.6	197	10.8	266	5.4
134	16.2	198	10.8	267	10.8
135	13.5	199	5.4	268	5.4
136	13.5	201	2.7	268	8.1
137	18.9	203	2.7	271	5.4
139	5.4	205	8.1	273	5.4
141	5.4	206	18.9	277	10.8
142	5.4	207	64.9	278	37.8
143	8.1	208	10.8	279	10.8
144	2.7	209	21.6	280	40.5
145	29.7	210	8.1	281	5.4
146	5.4	211	5.4	282	32.4
147	16.2	214	2.7	283	13.5
148	8.1	215	2.7	284	8.1
149	13.5	216	5.4	285	2.7
150	8.1	217	10.8	286	8.1
151	8.1	218	5.4	289	2.7
155	8.1	219	2.7	290	2.7
156	10.8	220	8.1	291	16.2
157	8.1	221	2.7	292	10.8
159	48.6	223	13.5	293	27.0
160	8.1	227	2.7	295	5.4
161	35.1	228	8.1	301	5.4
162	5.4	229	13.5	303	10.8
163	5.4	231	21.6	304	5.4
164	13.5	233	16.2	305	8.1
165	2.7	235	5.4	306	2.7
167	5.4	236	2.7	308	5.4
168	2.7	237	2.7	310	5.4
172	10.8	238	5.4	312	2.7
173	5.4	241	8.1	313	5.4
174	5.4	243	100.0	315	2.7

Peak 5–*continued*

Mass	Abundance	Mass	Abundance	Mass	Abundance
316	2.7	342	5.4	375	5.4
325	13.5	345	2.7	376	2.7
327	10.8	355	2.7	377	2.7
328	2.7	359	8.1	399	8.1
329	8.1	363	5.4	410	5.4
331	2.7	367	2.7	411	5.4
333	2.7	371	2.7	412	8.1
335	5.4	372	2.7	414	5.4
337	5.4	373	8.1	415	2.7
338	5.4	374	2.7	416	2.7
341	8.1				

Pesticide

CAS Name: Toxaphene

Synonyms

- Agricide maggot killer
- Alltex
- Alltox
- Camphechlor
- Camphochlor
- Chem-Phene
- Chlor Chem T-590
- Chlorinated camphene
- Compound 3956
- Ent 9,735
- Estonox
- Fasco-terpene
- Geniphene
- Gy-phene
- Hercules 3956
- Kamfochlor
- M 5055
- Melipax
- Motox
- NCI-C00259
- Octachlorocamphene
- PChK
- Penphene
- Phenacide
- Phenatox
- Polychlorcamphene
- Polychlorocamphene
- Strobane-T
- Synthetic 3956
- Toxadust
- Toxakil
- Toxyphen

CAS Registry No.: 8001-35-2 (formerly 8022-04-6, 12687-42-2, 12698-98-5, 12770-20-6, 37226-11-2)

ROTECS Ref.: XW52500

Merck Index Ref.: 9252

Major Ions

Peak 1: 159, 125, 197, 51, 75, 161, 209, 233

Peak 2: 100, 99, 109, 102, 197, 199, 125, 77

Peak 3: 159, 161, 125, 85, 75, 195, 83, 197

Peak 4: 85, 83, 100, 73, 329, 159, 51, 75

Peak 5: 245, 243, 83, 85, 207, 159, 73, 111

EPA Ions: 231, 233, 235

Selected Bibliography

The mass spectrometer as a substance-selective detector in chromatography, Budde, W. L., Eichelberger, J. W., *J. Chromatogr.*, **134**(1), 147–158 (1977).

A contribution to the composition of the insecticide "Toxaphene." Gas chromatographic–mass spectroscopic characterization of the oil and crystalline fraction of industrial Toxaphene and their insecticidal activity, Parlar, H., Nitz, S., Michna, A., Korte, F., *Z. Naturforsch., B: Anorg. Chem., Org. Chem.*, **33B**(8), 915–923 (1978).

TRICHLOROETHYLENE

This compound is a widely used solvent for cellulose ester and ethers, fats, oils, paints, resins, rubber, varnishes, waxes, etc. It is also employed in dry cleaning and in the manufacture of organic chemicals. It has been used as an insect fumigant, an analgesic, an antiseptic, and, in veterinary medicine, as an anesthetic.

TRICHLOROETHYLENE C_2HCl_3 (130)

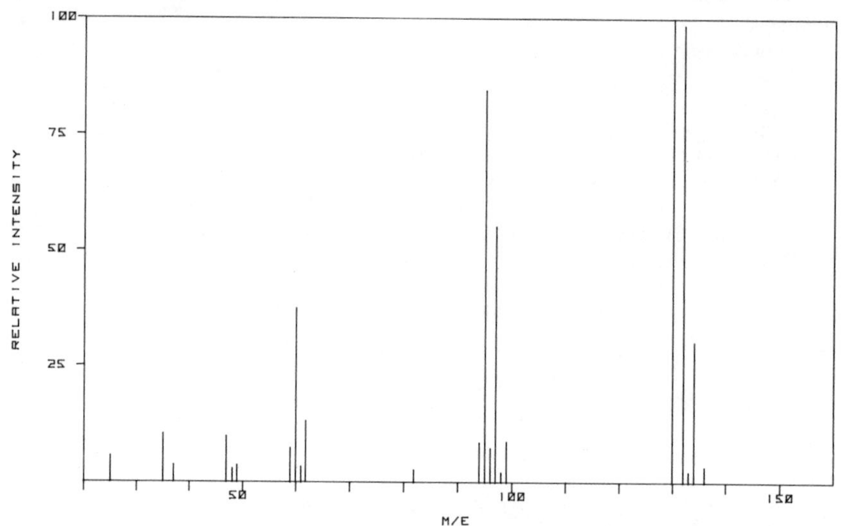

Spectral Data

Mass	Abundance	Mass	Abundance	Mass	Abundance
25	5.5	61	3.2	98	2.0
35	10.2	62	13.1	99	8.6
37	3.6	82	2.6	130	100.0
47	9.8	94	8.4	132	98.5
48	2.8	95	84.3	133	2.2
49	3.5	96	7.2	134	30.2
59	7.3	97	55.1	136	3.3
60	37.4				

TRICHLOROETHYLENE

Volatile
CAS Name: Ethylene, trichloro-
Synonyms

Acetylene trichloride
Algylen
Anamenth
Benzinol
Blacosolv
Blancosolv
Cecolene
Chlorilen
1-Chloro-2,2-dichloroethylene
Chlorylen
Chorylen
Circosolv
Crawhaspol
Densinfluat
1,1-Dichloro-2-chloroethylene
Dow-Tri
Dukeron
Ethinyl trichloride
Ethylene trichloride
Fleck-Flip
Flock-Flip
Fluate
Gemalgene
Germalgene
Lanadin
Lethurin
Narcogen
Narkogen
Narkosoid
NCI-C04546
Nialk
Perm-a-chlor
Perm-a-clor
Petzinol
Philex
TCE
Threthylen
Threthylene
Trethylene
Tri
Triad
Trial
Triasol
Trichloran
Trichloren
Trichlorethylene
Trichloroethene
1,1,2-Trichloroethylene
1,2,2-Trichloroethylene
Tri-Clene
Trielene
Trielin
Triklone
Trilen
Trilene
Triline
Trimar
Triol
Vestrol
Vitran
Westrosol

CAS Registry No.: 79-01-6
ROTECS Ref.: KX45500
Merck Index Ref.: 9319

Major Ions: 130, 132, 95, 97, 60, 134, 62, 35
EPA Ions: 95, 97, 130, 132

Selected Bibliography

Direct aqueous injection gas chromatography–mass spectrometry for analysis of organohalides in water at concentrations below the parts per billion level, Fujii, T., *J. Chromatogr.*, **139**(2), 297–302 (1977).

VINYL CHLORIDE

Vinyl chloride is used in the manufacture of polyvinyl chloride plastics, and was formerly employed as a refrigerant. Its use in the U.S. has been stringently regulated since workers handling the compound were found to have high incidences of angiosarcoma.

VINYL CHLORIDE C_2H_3Cl (62)

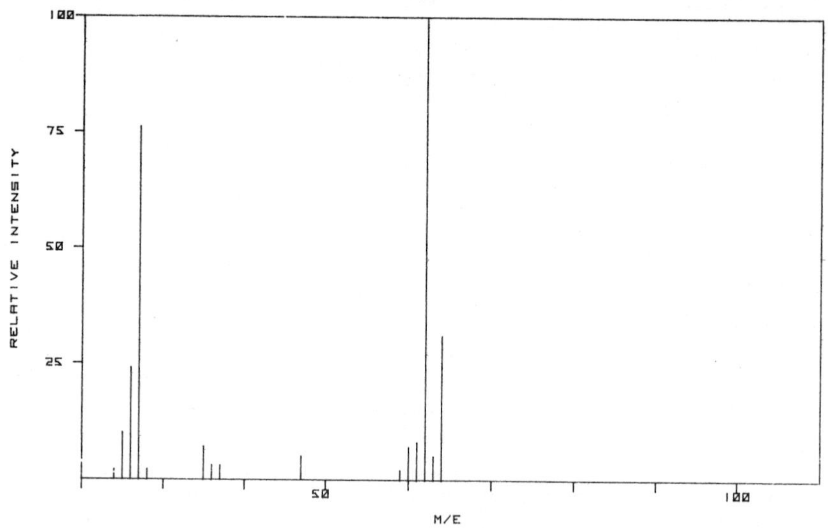

Spectral Data

Mass	Abundance	Mass	Abundance	Mass	Abundance
24	2.4	36	2.7	60	6.5
25	10.3	37	2.6	61	8.4
26	24.2	47	4.8	62	100.0
27	76.4	48	2.0	63	5.0
28	2.3	49	1.7	64	30.8
35	7.3	59	2.0		

VINYL CHLORIDE

Volatile
CAS Name: Ethylene, chloro-
Synonyms

Chlorethene	Monochloroethylene
Chlorethylene	Trovidur
Chloroethene	VC
Chloroethylene	VCM
Ethylene monochloride	Vinyl chloride monomer
Monochloroethene	Vinyl C monomer

CAS Registry No.: 75-01-4
ROTECS Ref.: KU96250
Merck Index Ref.: 9645

Major Ions: 62, 27, 64, 26, 25, 61, 35, 60
EPA Ions: 62, 64

Selected Bibliography

Analysis of volatile compounds in plastic products. I. Microanalysis of vinyl chloride monomer in poly(vinyl chloride) products by mass fragmentography, Baba, T., *Shokuhin Eiseigaku Zasshi*, 18(6), 500–503 (1977).

Analysis of volatile compounds in plastic products. II. Extraction method by solvent vapor for the analysis of vinyl chloride monomer in poly(vinyl chloride), Baba, T., *Shokuhin Eiseigaku Zasshi*, 18(6), 504–508 (1977).

Trace determination of vinyl chloride in water by direct aqueous injection gas chromatography–mass spectrometry, Fujii, T., *Anal. Chem.*, 49(13), 1985–1987 (1977).

The mass spectrometer as a continuous monitor for pollutants and toxic hazards, Ball, G. W., Parker, R. B., Stevens, K., *Adv. Mass Spectrom.*, 7B, 1687–1690 (1978).

INORGANIC SUBSTANCES

Antimony, arsenic, asbestos, beryllium, cadmium, chromium, copper, cyanide, lead, mercury, nickel, selenium, silver, thallium, and zinc are the substances included in this category.

The E.P.A. protocol requires that the metals be analyzed by atomic absorption spectroscopy. Mass spectrometry has been used for the analysis of many individual organometallic compounds, but it is inappropriate to tabulate their spectral data here. For a series of comprehensive reviews, see the "selected bibliography" in this section.

ANTIMONY Sb (121)

Inorganic
CAS Name: Antimony
Synonyms
- Antimony black
- Antimony, regulus
- C. I. 77050
- Regulus of antimony
- Stibium

CAS Registry No.: 7440-36-0
ROTECS Ref.: CC40250
Merck Index Ref.: 729

ARSENIC As (75)

Inorganic
CAS Name: Arsenic
Synonyms
- Arsenic black
- Colloidal arsenic
- Grey arsenic

CAS Registry No.: 7440-38-2
ROTECS Ref.: CG05250
Merck Index Ref.: 820

ASBESTOS (mixture)

Inorganic
CAS Name (1): Anthophyllite
Synonyms (1)
16F
Azbolen asbestos

CAS Registry No. (1): 17068-78-9 (formerly 37229-03-1)

CAS Name (2): Chrysotile
Synonyms (2)

Avibest C	Cassiar AK	Plastibest 20
Calidria RG 100	Chrysotile asbestos	5R04
Calidria RG 144	Hooker No. 1 chrysotile asbestos	RG600
Calidria RG 600	K 6–30	Serpentine chrysotile

CAS Registry No. (2): 12001-29-5

CAS Name (3): Asbestos
Synonym (3)
Asbestos Fiber

CAS Registry No. (3): 1332-21-4 (formerly 12413-45-5)
ROTECS Ref. (3): CI64750
Merck Index Ref. (3): 850

BERYLLIUM Be (9)

Inorganic
CAS Name: Beryllium
Synonym
Glucinium

CAS Registry No.: 7440-14-7
ROTECS Ref.: DS17500
Merck Index Ref.: 1184

CADMIUM Cd (114)

Inorganic
CAS Name: Cadmium
Synonym
C.I. 77180

CAS Registry No.: 7440-43-9
ROTECS Ref.: EU98000
Merck Index Ref.: 1600

CHROMIUM — Cr (52)

Inorganic
CAS Name: Chromium
Synonym
Chrome

CAS Registry No.: 7440-47-3
ROTECS Ref.: GB42000
Merck Index Ref.: 2229

COPPER — Cu (63)

Inorganic
CAS Name: Copper
Synonyms

1721 Gold	C.I. pigment metal 2	LCu
Allbri natural copper	Copper M1	M1
Anac 110	Copper bronze	M2
Arwood copper	Copper powder	M3
Bronze powder	CuEP	M3R
CDA 101	CuEPP	M3S
CDA 102	DCuP1	M4
CDA 110	Gold bronze	OFHC Cu
CDA 122	Kafar copper	Raney copper
C.I. 77400		

CAS Registry No.: 7440-50-8 (formerly 12711-87-4)
ROTECS Ref.: GL53250
Merck Index Ref.: 2496

CYANIDE — CN^- (26)

Inorganic
CAS Name: Cyanide
Synonyms
Carbon nitride ion
Cyanide anion
Cyanide ion
Hydrocyanic acid ion
Isocyanide

CAS Registry No.: 57-12-5
ROTECS Ref.: GS71750

LEAD Pb (208)

Inorganic
CAS Name: Lead
Synonyms

C.I. 77575	Lead S2
C.I. pigment metal 4	S0
KS-4	S1
Lead Flake	

CAS Registry No.: 7439-92-1
ROTECS Ref.: OF75250
Merck Index Ref.: 5242

MERCURY Hg (202)

Inorganic
CAS Name: Mercury
Synonyms

Hydrargyrum
Liquid silver
Quicksilver

CAS Registry No.: 7439-97-6 (formerly 8030-64-6, 51887-47-9)
ROTECS Ref.: OV45500
Merck Index Ref.: 5742

NICKEL Ni (58)

Inorganic
CAS Name: Nickel
Synonyms

Carbonyl nickel powder	Nickel catalyst, wet
Ni 270	Nickel sponge
Ni 0901-S (Harshaw)	Pulverized nickel
Ni 4303-T	Raney nickel
Nickel 270	RCH 55/5

CAS Registry No.: 7440-02-0 (formerly 8049-31-8, 17375-04-1, 39303-46-3, 53527-81-4)
ROTECS Ref.: QR59500
Merck Index Ref.: 6312

SELENIUM — Se (80)

Inorganic
CAS Name: Selenium
Synonyms
- C.I. 77805
- Elemental selenium
- Selenium dust
- Selenium homopolymer

CAS Registry No.: 7782-49-2 (formerly 11125-23-8, 11133-88-3, 12640-29-8, 12640-30-1, 12641-96-2, 12733-65-2, 37256-19-2, 37258-85-8, 37276-15-6, 37368-02-8)
ROTECS Ref.: VS77000
Merck Index Ref.: 8179

SILVER — Ag (107)

Inorganic
CAS Name: Silver
Synonyms

Argentum	Silflake 135
C.I. 77820	Silver metal
L-3	Sr 999
Shell silver	V9

CAS Registry No.: 7440-22-4
ROTECS Ref.: VW35000
Merck Index Ref.: 8244

THALLIUM — Tl (205)

Inorganic
CAS Name: Thallium
Synonym
- Ramor

CAS Registry No.: 7440-28-0
Merck Index Ref.: 8970

ZINC Zn (64)

Inorganic
CAS Name: Zinc
Synonym

Blue powder	Granular zinc
C.I. 77945	Jasad
C.I. pigment black 16	Zinc dust
C.I. pigment metal 6	Zinc powder
Emanay zinc dust	

CAS Registry No.: 7440-66-6 (formerly 12793-53-2)
ROTECS Ref.: ZG86000
Merck Index Ref.: 9782

Selected Bibliography

Organometallic, Co-ordination, and Inorganic Compounds, Spalding, T. R., *Specialist Periodical Report: Mass Spectrometry*, **5**, 312 (1979).

Organometallic, Coordination, and Inorganic Compounds, Spalding, T. R., *Specialist Periodical Report: Mass Spectrometry*, **4**, 268 (1977).

Organometallic, Co-ordination, and Inorganic Compounds, Spalding, T. R., *Specialist Periodical Report: Mass Spectrometry*, **3**, 143 (1975).

Organometallic and Co-ordination Compounds, Bruce, M. I., *Specialist Periodical Report: Mass Spectrometry*, **2**, 193 (1973).

Organometallic and Co-ordination Compounds, Bruce, M. I., *Specialist Periodical Report: Mass Spectrometry*, **1**, 182 (1971).

INTERNAL STANDARDS

Anthracene-d_{10} is used for nonvolatiles, while bromochloromethane, 2-bromo-1-chloropropane, and 1,4-dichlorobutane are employed for volatiles.

ANTHRACENE-d_{10} $C_{14}D_{10}$ (188)

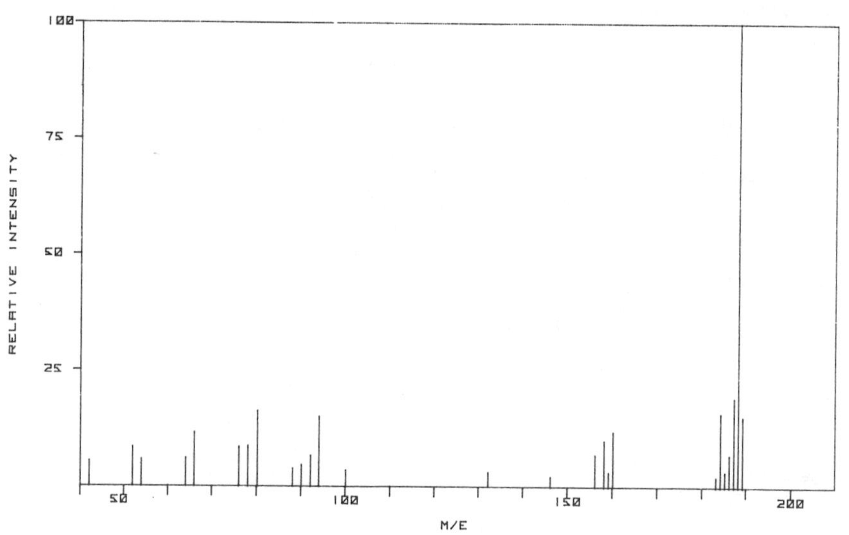

Spectral Data

Mass	Abundance	Mass	Abundance	Mass	Abundance
42	5.3	90	4.5	160	11.8
52	8.3	92	6.5	183	2.1
54	5.7	94	14.9	184	15.7
64	6.0	100	3.4	185	3.2
66	11.5	132	3.1	186	6.8
76	8.3	146	2.2	187	19.1
78	8.5	156	6.8	188	100.0
80	16.1	158	9.8	189	15.0
88	3.7	159	3.1		

Internal standard (extractables)
Synonyms
[$^2H_{10}$]Anthracene
Perdeuterioanthracene
Perdeuteroanthracene

CAS Registry No.: 1719-06-8

Major Ions: 188, 187, 80, 184, 189, 94, 160, 66
EPA Ions: 188, 94, 80

BROMOCHLOROMETHANE — CH_2BrCl (128)

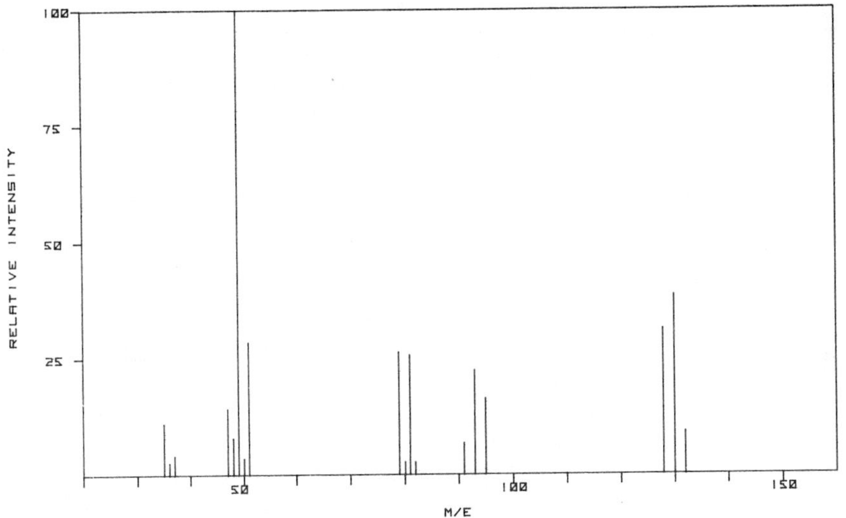

Spectral Data

Mass	Abundance	Mass	Abundance	Mass	Abundance
35	10.9	50	3.4	91	6.6
36	2.5	51	28.4	93	22.4
37	4.0	79	26.2	95	16.2
47	14.0	80	2.6	128	31.1
48	7.7	81	25.6	130	38.4
49	100.0	82	2.6	132	9.0

BROMOCHLOROMETHANE–*continued*

Internal standard (volatiles)
CAS Name: Methane, bromochloro-
Synonyms
Chlorobromomethane
Halon 1011
Methylene chlorobromide
MIL-B-4394-B
Monochloromonobromomethane

CAS Registry No.: 74-97-5
ROTECS Ref.: PA52500

Major Ions: 49, 130, 128, 51, 79, 81, 93, 95
EPA Ions: 49, 130, 128, 51

2-BROMO-1-CHLOROPROPANE — C_3H_6BrCl (156)

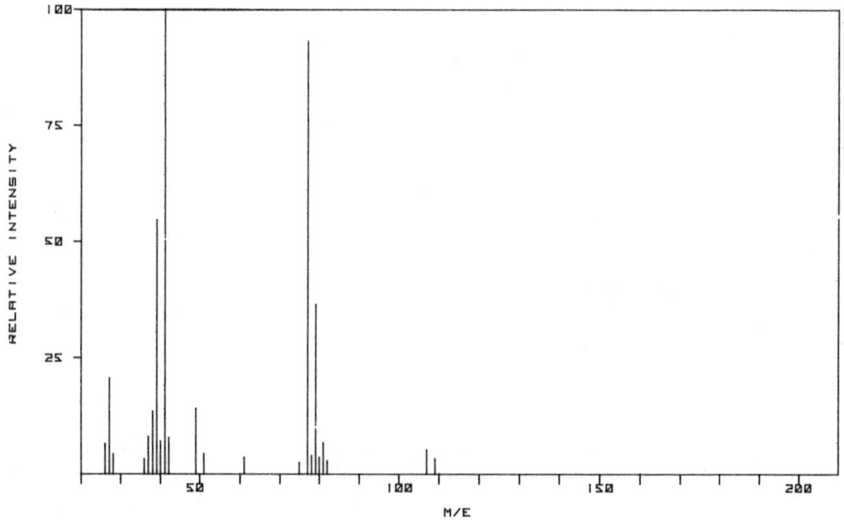

Spectral Data

Mass	Abundance	Mass	Abundance	Mass	Abundance
26	6.4	41	100.0	78	3.9
27	20.5	42	7.7	79	36.4
28	4.2	49	14.0	80	3.5
36	3.1	51	4.3	81	6.6
37	7.9	61	3.5	82	2.8
38	13.4	75	2.4	107	5.1
39	54.6	77	93.2	109	3.2
40	6.9				

Internal standard (volatiles)
CAS Name: Propane, 2-bromo-1-chloro-
Synonym
1-Chloro-2-bromopropane

CAS Registry No.: 3017-95-6

Major Ions: 41, 77, 39, 79, 27, 49, 38, 37
EPA Ions: 77, 79, 156

1,4-DICHLOROBUTANE $C_4H_8Cl_2$ (126)

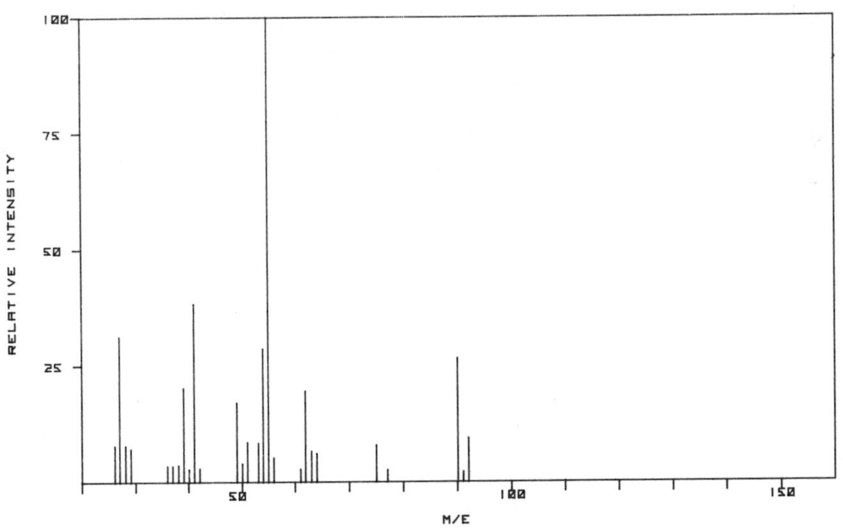

Spectral Data

Mass	Abundance	Mass	Abundance	Mass	Abundance
26	7.7	41	38.2	61	2.6
27	31.2	42	2.8	62	19.5
28	7.7	49	17.0	63	6.5
29	7.0	50	3.9	64	6.0
36	3.3	51	8.4	75	7.7
37	3.3	53	8.2	77	2.5
38	3.5	54	28.6	90	26.5
39	20.2	55	100.0	91	2.1
40	2.6	56	5.1	92	9.3

Internal standard (volatiles)
CAS Name: Butane, 1,4-dichloro-
CAS Registry No.: 110-56-5

Major Ions: 55, 41, 27, 54, 90, 39, 62, 49
EPA Ions: 55, 90, 92

APPENDIX

Each of the indexes in this appendix is arranged according to the mass numbers given in the shaded columns, i.e., the column under *M* in the Molecular Weight Index, and the first, second, and third columns, respectively, in the Base-, Second-, and Third-Peak Indexes. In each index, the next to last column (*M*) gives the molecular weight.

Molecular Weight Index

Partial spectrum								M	Substance
50 (100.0)	52 (34.2)	49 (13.3)	47 (11.9)	35 (8.4)	48 (5.0)	51 (4.6)	37 (3.0)	**50** (100.0)	Chloromethane
53 (100.0)	52 (79.6)	26 (78.5)	51 (33.5)	27 (16.0)	25 (10.2)	38 (9.8)	50 (9.1)	**53** (100.0)	Acrylonitrile
56 (100.0)	27 (66.2)	55 (64.9)	26 (44.5)	29 (37.7)	37 (11.3)	38 (9.3)	53 (8.6)	**56** (100.0)	Acrolein
62 (100.0)	27 (76.4)	64 (30.8)	26 (24.2)	25 (10.3)	61 (8.4)	35 (7.3)	60 (6.5)	**62** (100.0)	Vinyl chloride
64 (100.0)	28 (84.9)	27 (82.6)	29 (78.1)	26 (36.9)	66 (35.0)	49 (27.0)	51 (8.8)	**64** (100.0)	Chloroethane
74 (100.0)	42 (54.1)	43 (24.9)	40 (3.8)	41 (3.1)	75 (3.0)	44 (2.8)	–	**74** (100.0)	*N*-Nitrosodimethylamine
78 (100.0)	77 (25.9)	50 (20.8)	51 (20.6)	52 (18.8)	39 (9.3)	79 (6.4)	74 (5.8)	**78** (100.0)	Benzene
49 (100.0)	84 (60.1)	86 (38.3)	51 (28.4)	47 (22.0)	35 (14.6)	48 (9.6)	28 (6.6)	**84** (60.1)	Methylene chloride
91 (100.0)	92 (57.8)	65 (14.4)	39 (12.8)	63 (10.4)	51 (8.3)	50 (5.9)	62 (5.0)	**92** (57.8)	Toluene
94 (100.0)	96 (89.3)	93 (21.6)	79 (21.3)	81 (20.5)	95 (14.1)	91 (9.0)	92 (4.3)	**94** (100.0)	Bromomethane
94 (100.0)	66 (32.3)	65 (22.0)	95 (6.6)	63 (5.6)	55 (4.7)	50 (3.9)	51 (3.5)	**94** (100.0)	Phenol
61 (100.0)	96 (44.3)	63 (31.9)	98 (28.9)	60 (18.9)	26 (13.4)	35 (9.2)	62 (8.4)	**96** (44.3)	1,1-Dichloroethylene

61 (100.0)	96 (60.2)	98 (39.3)	63 (32.3)	60 (29.7)	26 (16.3)	62 (12.2)	25 (8.6)	**96** (60.2)	*trans*-1,2-Dichloroethylene
63 (100.0)	27 (42.8)	65 (31.4)	26 (13.2)	83 (12.9)	61 (8.9)	62 (8.4)	85 (8.3)	**98** (6.8)	1,1-Dichloroethane
62 (100.0)	27 (39.8)	64 (31.1)	49 (24.9)	26 (13.1)	63 (12.5)	61 (8.8)	51 (7.8)	**98** (7.3)	1,2-Dichloroethane
63 (100.0)	43 (98.1)	44 (71.5)	65 (32.3)	42 (25.0)	62 (24.7)	106 (24.5)	57 (20.2)	**106** (24.5)	2-Chloroethyl vinyl ether
91 (100.0)	106 (27.3)	65 (10.9)	51 (10.7)	77 (9.2)	78 (8.0)	92 (7.6)	39 (6.5)	**106** (27.3)	Ethylbenzene
75 (100.0)	39 (50.3)	77 (28.9)	49 (21.2)	110 (19.6)	38 (13.9)	112 (12.2)	37 (12.1)	**110** (19.6)	*cis*-1,3-Dichloropropene
75 (100.0)	39 (55.0)	77 (29.1)	49 (20.4)	110 (19.8)	38 (14.4)	37 (12.8)	112 (11.7)	**110** (19.8)	*trans*-1,3-Dichloropropene
112 (100.0)	77 (67.6)	114 (31.4)	51 (25.8)	50 (25.0)	45 (11.9)	74 (11.4)	75 (9.7)	**112** (100.0)	Chlorobenzene
63 (100.0)	62 (79.2)	41 (70.1)	76 (61.7)	39 (54.9)	27 (44.7)	65 (30.8)	64 (27.0)	**112** (5.8)	1,2-Dichloropropane
79 (100.0)	49 (37.9)	81 (33.1)	51 (11.9)	29 (9.5)	78 (5.5)	80 (3.8)	48 (2.9)	**114**	Bis(chloromethyl) ether
83 (100.0)	85 (64.7)	47 (29.7)	35 (14.4)	48 (11.9)	49 (10.5)	87 (9.5)	37 (4.8)	**118**	Chloroform
85 (100.0)	87 (30.9)	50 (24.6)	35 (22.0)	31 (16.3)	47 (8.7)	101 (7.6)	66 (6.8)	**120**	Dichlorodifluoromethane
107 (100.0)	122 (92.1)	121 (49.6)	77 (26.9)	91 (19.4)	79 (15.0)	78 (9.0)	51 (8.9)	**122** (92.1)	2,4-Dimethylphenol

Molecular Weight Index—continued

Partial spectrum								M	Substance
77 (100.0)	123 (75.6)	51 (43.8)	50 (16.0)	93 (14.8)	65 (14.0)	74 (7.9)	78 (6.7)	**123** (75.6)	Nitrobenzene
55 (100.0)	41 (38.2)	27 (31.2)	54 (28.6)	90 (26.5)	39 (20.2)	62 (19.5)	49 (17.0)	**126**	1,4-Dichlorobutane (IS)
49 (100.0)	130 (38.4)	128 (31.1)	51 (28.4)	79 (26.2)	81 (25.6)	93 (22.4)	95 (16.2)	**128** (31.1)	Bromochloromethane (IS)
128 (100.0)	130 (31.2)	64 (29.3)	63 (16.1)	92 (12.3)	129 (6.7)	73 (5.0)	65 (4.8)	**128** (100.0)	2-Chlorophenol
128 (100.0)	127 (14.0)	129 (11.4)	102 (8.9)	126 (7.8)	51 (5.8)	75 (4.9)	63 (4.7)	**128** (100.0)	Naphthalene
70 (100.0)	43 (70.7)	130 (44.8)	42 (41.4)	41 (33.9)	101 (16.6)	58 (15.8)	113 (10.8)	**130** (44.8)	*N*-Nitrosodi-n-propylamine
130 (100.0)	132 (98.5)	95 (84.3)	97 (55.1)	60 (37.4)	134 (30.2)	62 (13.1)	35 (10.2)	**130** (100.0)	Trichloroethylene
97 (100.0)	99 (60.9)	61 (51.2)	63 (16.3)	26 (11.8)	117 (10.7)	119 (10.2)	101 (10.2)	**132**	1,1,1-Trichloroethane
97 (100.0)	83 (76.0)	99 (61.7)	61 (54.7)	85 (53.0)	96 (28.3)	98 (20.6)	63 (17.7)	**132** (15.7)	1,1,2-Trichloroethane
101 (100.0)	103 (59.1)	66 (17.6)	35 (12.8)	105 (11.0)	47 (10.9)	31 (9.3)	68 (6.0)	**136**	Trichlorofluoromethane
82 (100.0)	54 (14.8)	138 (12.2)	41 (7.7)	53 (6.9)	83 (5.3)	67 (5.3)	55 (5.3)	**138** (12.2)	Isophorone
139 (100.0)	65 (68.0)	63 (59.2)	81 (45.2)	53 (34.8)	64 (34.2)	109 (26.7)	51 (19.0)	**139** (100.0)	2-Nitrophenol

65 (100.0)	139 (83.6)	109 (43.4)	53 (29.7)	81 (27.6)	63 (27.3)	93 (18.9)	62 (14.8)	**139 (83.6)**	4-Nitrophenol
93 (100.0)	63 (57.8)	95 (34.4)	65 (16.6)	49 (5.4)	94 (3.8)	106 (3.0)	142 (2.9)	**142 (2.9)**	Bis(2-chloroethyl) ether
107 (100.0)	142 (76.4)	77 (43.7)	144 (24.8)	51 (17.4)	78 (12.1)	79 (11.8)	143 (9.0)	**142 (76.4)**	*p*-Chloro-*m*-cresol
146 (100.0)	148 (60.1)	75 (52.5)	111 (51.5)	50 (37.5)	74 (31.8)	113 (18.0)	73 (15.9)	**146 (100.0)**	1,2-Dichlorobenzene
146 (100.0)	148 (58.3)	75 (50.6)	111 (49.7)	50 (36.5)	74 (32.4)	73 (18.1)	113 (16.0)	**146 (100.0)**	1,3-Dichlorobenzene
146 (100.0)	148 (58.2)	111 (33.4)	75 (20.8)	74 (13.0)	113 (11.0)	150 (10.1)	50 (9.4)	**146 (100.0)**	1,4-Dichlorobenzene
152 (100.0)	151 (17.5)	150 (13.5)	153 (12.6)	76 (9.3)	75 (3.8)	74 (3.7)	126 (3.6)	**152 (100.0)**	Acenaphthylene
117 (100.0)	119 (93.7)	121 (30.7)	82 (28.3)	47 (25.6)	35 (18.7)	84 (18.3)	49 (8.9)	**152**	Carbon tetrachloride
153 (100.0)	154 (96.5)	152 (44.6)	76 (19.0)	151 (16.7)	155 (14.0)	63 (8.9)	150 (8.5)	**154 (96.5)**	Acenaphthene
41 (100.0)	77 (93.2)	39 (54.6)	79 (36.4)	27 (20.5)	49 (14.0)	38 (13.4)	37 (7.9)	**156**	2-Bromo-1-chloropropane (IS)
83 (100.0)	85 (66.5)	47 (20.8)	48 (12.8)	79 (12.4)	81 (11.9)	129 (10.1)	87 (9.4)	**162**	Bromodichloromethane
162 (100.0)	127 (32.1)	164 (29.0)	126 (17.3)	163 (10.9)	63 (8.3)	75 (7.6)	81 (6.7)	**162 (100.0)**	2-Chloronaphthalene
162 (100.0)	164 (62.2)	63 (45.4)	98 (31.5)	126 (13.9)	99 (12.3)	62 (12.2)	73 (11.0)	**162 (100.0)**	2,4-Dichlorophenol
166 (100.0)	129 (77.1)	164 (74.2)	131 (73.4)	168 (54.3)	94 (53.8)	47 (35.7)	96 (33.6)	**164 (74.2)**	1,1,2,2-Tetrachloroethene

Molecular Weight Index—continued

Partial spectrum								*M*	Substance
166 (100.0)	165 (90.9)	163 (14.2)	164 (12.6)	167 (12.4)	139 (6.5)	115 (4.2)	63 (3.9)	**166** (100.0)	Fluorene
83 (100.0)	85 (66.3)	95 (15.6)	60 (14.1)	61 (11.9)	96 (11.1)	168 (9.9)	87 (9.8)	**166** (7.7)	1,1,2,2-Tetrachloroethane
45 (100.0)	121 (69.1)	41 (64.1)	77 (26.4)	123 (24.0)	79 (18.3)	49 (11.1)	107 (10.7)	**170**	Bis(2-chloroisopropyl) ether
93 (100.0)	63 (68.7)	95 (30.4)	123 (24.9)	65 (22.1)	125 (7.2)	106 (4.9)	49 (4.5)	**172**	Bis(2-chloroethoxy)methane
178 (100.0)	176 (17.7)	179 (13.9)	177 (8.6)	76 (8.2)	89 (7.6)	152 (7.2)	151 (5.6)	**178** (100.0)	Anthracene
178 (100.0)	176 (17.9)	179 (14.2)	76 (10.3)	177 (10.1)	152 (9.4)	89 (8.8)	88 (7.3)	**178** (100.0)	Phenanthrene
180 (100.0)	182 (96.3)	184 (30.2)	145 (29.0)	109 (22.1)	74 (21.6)	147 (17.1)	75 (12.8)	**180** (100.0)	1,2,4-Trichlorobenzene
165 (100.0)	89 (64.4)	63 (53.0)	90 (22.7)	119 (18.5)	78 (18.4)	51 (18.1)	64 (17.2)	**182** (9.1)	2,4-Dinitrotoluene
165 (100.0)	63 (47.7)	89 (41.7)	90 (30.4)	77 (23.9)	51 (22.2)	64 (21.0)	78 (20.8)	**182** (2.5)	2,6-Dinitrotoluene
184 (100.0)	185 (12.3)	183 (11.5)	92 (6.8)	156 (6.4)	91 (5.9)	167 (5.1)	166 (4.9)	**184** (100.0)	Benzidine
184 (100.0)	63 (67.9)	53 (61.2)	91 (39.3)	62 (35.7)	107 (33.9)	154 (31.3)	79 (31.3)	**184** (100.0)	2,4-Dinitrophenol
77 (100.0)	51 (37.6)	182 (24.2)	105 (15.5)	78 (11.3)	50 (9.5)	152 (5.7)	153 (3.4)	**184**	1,2-Diphenylhydrazine [azobenzene]

188 (100.0)	187 (19.1)	80 (16.1)	184 (15.7)	189 (15.0)	94 (14.9)	160 (11.8)	66 (11.5)	**188 (100.0)**	Anthracene-d_{10} (IS)
163 (100.0)	77 (17.7)	194 (10.2)	164 (9.5)	92 (8.0)	76 (7.6)	50 (7.2)	133 (6.2)	**194 (10.2)**	Dimethyl phthalate
196 (100.0)	198 (97.5)	97 (56.8)	132 (42.7)	200 (33.9)	134 (28.0)	62 (25.7)	99 (22.9)	**196 (100.0)**	2,4,6-Trichlorophenol
198 (100.0)	51 (52.3)	105 (44.5)	53 (41.9)	121 (41.5)	77 (24.7)	52 (24.2)	50 (24.2)	**198 (100.0)**	4,6-Dinitro-*o*-cresol
77 (100.0)	51 (23.7)	182 (21.9)	105 (18.4)	78 (7.3)	152 (5.3)	50 (4.6)	183 (2.9)	**198**	*N*-Nitrosodiphenylamine [diphenylamine]
202 (100.0)	200 (20.2)	203 (18.8)	101 (15.9)	201 (13.9)	100 (11.3)	88 (8.7)	87 (5.2)	**202 (100.0)**	Fluoranthene
202 (100.0)	200 (22.2)	201 (19.7)	203 (19.5)	101 (15.4)	100 (13.0)	88 (5.5)	199 (4.5)	**202 (100.0)**	Pyrene
204 (100.0)	141 (50.0)	206 (36.9)	77 (28.7)	51 (20.3)	115 (12.7)	75 (12.7)	205 (12.0)	**204 (100.0)**	4-Chlorophenyl phenyl ether
129 (100.0)	127 (76.2)	97 (39.6)	83 (36.0)	61 (35.2)	79 (24.1)	131 (23.7)	81 (23.4)	**206 (2.6)**	Dibromochloromethane
149 (100.0)	177 (23.7)	65 (10.6)	176 (10.5)	150 (10.2)	76 (9.6)	105 (8.5)	121 (7.0)	**222 (4.6)**	Diethyl phthalate
228 (100.0)	226 (26.5)	113 (23.0)	114 (22.8)	229 (20.6)	101 (12.5)	112 (10.0)	100 (9.5)	**228 (100.0)**	Benzo[*a*]anthracene
228 (100.0)	226 (30.9)	113 (26.3)	114 (22.2)	229 (20.3)	227 (12.5)	112 (11.0)	100 (10.3)	**228 (100.0)**	Chrysene
117 (100.0)	119 (83.0)	201 (57.8)	203 (37.5)	199 (34.7)	166 (32.3)	94 (30.5)	47 (30.4)	**234**	Hexachloroethane
248 (100.0)	250 (88.6)	141 (69.5)	77 (58.5)	51 (37.3)	115 (29.3)	50 (24.4)	63 (22.5)	**248 (100.0)**	4-Bromophenyl phenyl ether

Molecular Weight Index—continued

Partial spectrum								*M*	Substance
173 (100.0)	171 (51.9)	175 (50.7)	79 (30.6)	91 (29.7)	93 (29.6)	81 (28.9)	94 (16.9)	**250** (3.3)	Bromoform
252 (100.0)	126 (23.0)	250 (21.7)	253 (20.5)	125 (14.7)	113 (10.8)	124 (9.7)	112 (8.8)	**252** (100.0)	Benzo[*b*]fluoranthene
252 (100.0)	250 (20.9)	253 (20.7)	126 (14.9)	125 (10.2)	113 (7.1)	112 (6.5)	251 (6.2)	**252** (100.0)	Benzo[*k*]fluoranthene
252 (100.0)	250 (22.1)	253 (22.0)	126 (16.9)	125 (10.4)	113 (9.7)	251 (8.2)	112 (6.8)	**252** (100.0)	Benzo[*a*]pyrene
252 (100.0)	254 (59.1)	253 (17.8)	126 (17.3)	127 (15.5)	154 (13.7)	91 (12.6)	181 (10.8)	**252** (100.0)	3,3′-Dichlorobenzidine
225 (100.0)	223 (67.2)	227 (64.1)	190 (45.9)	188 (37.9)	118 (34.8)	260 (33.6)	262 (23.6)	**258** (20.8)	Hexachlorobutadiene
266 (100.0)	268 (74.3)	264 (70.0)	165 (45.5)	95 (39.5)	167 (38.7)	60 (26.5)	270 (26.1)	**264** (70.0)	Pentachlorophenol
237 (100.0)	235 (73.0)	239 (69.7)	95 (28.4)	130 (22.6)	241 (22.2)	272 (17.8)	60 (16.9)	**270** (9.7)	Hexachlorocyclopentadiene
276 (100.0)	277 (23.7)	274 (21.3)	138 (19.4)	137 (17.8)	275 (11.4)	136 (8.9)	272 (5.2)	**276** (100.0)	Benzo[*ghi*]perylene
276 (100.0)	277 (24.1)	274 (21.5)	138 (18.2)	137 (13.9)	275 (9.2)	136 (7.8)	125 (5.2)	**276** (100.0)	Indeno[1,2,3-*cd*]pyrene
278 (100.0)	139 (38.6)	138 (27.2)	279 (23.4)	276 (21.9)	137 (16.3)	125 (15.2)	113 (8.8)	**278** (100.0)	Dibenzo[*a*,*h*]anthracene
149 (100.0)	150 (9.0)	41 (6.3)	223 (5.7)	104 (5.0)	76 (4.8)	205 (4.5)	65 (3.4)	**278**	Di-n-butyl phthalate

284 (100.0)	286 (83.7)	282 (50.7)	142 (35.3)	288 (32.8)	249 (28.1)	107 (25.6)	144 (22.6)	**282** (50.7)	Hexachlorobenzene
181 (100.0)	183 (91.7)	219 (84.7)	217 (66.0)	109 (65.3)	111 (62.4)	51 (43.0)	85 (36.9)	**288**	α-BHC
109 (100.0)	181 (89.3)	183 (82.7)	111 (82.4)	217 (65.5)	51 (48.8)	85 (46.0)	83 (38.6)	**288**	β-BHC
181 (100.0)	183 (98.5)	111 (96.9)	219 (83.0)	109 (74.4)	217 (58.3)	51 (55.6)	73 (55.4)	**288** (2.9)	γ-BHC
181 (100.0)	183 (94.0)	219 (89.6)	109 (85.0)	111 (74.9)	217 (64.5)	51 (55.2)	85 (49.7)	**288**	δ-BHC
149 (100.0)	91 (69.5)	206 (24.4)	65 (16.4)	123 (14.9)	104 (14.2)	132 (12.7)	150 (11.2)	**312** (2.0)	Butyl benzyl phthalate
246 (100.0)	318 (76.8)	248 (72.9)	316 (57.1)	176 (57.0)	105 (40.4)	320 (35.0)	75 (23.8)	**316** (57.1)	4,4′-DDE
235 (100.0)	165 (58.4)	237 (57.0)	75 (21.2)	199 (13.5)	88 (12.3)	236 (12.2)	82 (10.2)	**318** (2.6)	4,4′-DDD
322 (100.0)	320 (78.5)	324 (47.9)	74 (37.4)	97 (36.0)	259 (34.3)	257 (32.9)	194 (31.6)	**320** (78.5)	2,3,7,8-Tetrachlorodibenzo-*p*-dioxin
235 (100.0)	237 (56.6)	165 (53.8)	75 (17.6)	176 (13.3)	236 (12.5)	199 (11.9)	212 (11.0)	**352**	4,4′-DDT
66 (100.0)	91 (42.7)	79 (38.9)	65 (31.9)	101 (29.6)	263 (29.4)	265 (22.3)	261 (22.3)	**362**	Aldrin
100 (100.0)	272 (52.8)	65 (47.8)	274 (45.0)	102 (34.1)	270 (28.5)	237 (21.6)	135 (17.7)	**370** (2.2)	Heptachlor
79 (100.0)	82 (34.8)	81 (34.7)	77 (22.6)	108 (15.4)	53 (13.6)	107 (11.5)	263 (10.1)	**378**	Dieldrin
67 (100.0)	83 (55.6)	317 (50.8)	85 (45.7)	250 (36.1)	345 (35.6)	319 (33.7)	315 (33.3)	**378** (2.8)	Endrin

Molecular Weight Index—continued

Partial spectrum								*M*	Substance
67 (100.0)	250 (40.0)	345 (30.0)	197 (30.0)	135 (30.0)	95 (29.0)	248 (27.1)	66 (26.7)	**378**	Endrin aldehyde
81 (100.0)	353 (38.6)	355 (35.9)	53 (25.0)	351 (24.6)	51 (20.7)	357 (16.2)	61 (15.7)	**386** (2.3)	Heptachlor epoxide
149 (100.0)	167 (43.1)	57 (23.3)	71 (15.4)	70 (15.0)	43 (14.3)	279 (13.7)	41 (13.6)	**390**	Bis(2-ethylhexyl) phthalate
149 (100.0)	279 (10.7)	150 (9.7)	43 (7.6)	41 (7.5)	57 (5.5)	71 (3.8)	69 (3.2)	**390**	Di-n-octyl phthalate
195 (100.0)	170 (79.3)	207 (75.2)	239 (71.0)	237 (69.0)	159 (67.6)	241 (66.9)	197 (62.8)	**404** (2.1)	α-Endosulfan
195 (100.0)	197 (80.5)	159 (75.6)	160 (65.9)	207 (63.4)	170 (53.7)	239 (51.2)	237 (51.2)	**404** (4.9)	β-Endosulfan
272 (100.0)	274 (97.7)	229 (71.6)	270 (62.5)	227 (62.5)	237 (56.8)	387 (48.9)	239 (48.9)	**420** (5.7)	Endosulfan sulfate
100 (100.0)	102 (35.2)	238 (33.0)	65 (30.4)	101 (23.7)	236 (19.3)	240 (17.4)	66 (14.4)		Chlordane (peak 1)
100 (100.0)	272 (54.9)	274 (44.7)	102 (38.1)	65 (35.0)	270 (33.2)	237 (20.4)	276 (17.3)		Chlordane (peak 2)
232 (100.0)	303 (94.8)	230 (94.8)	169 (75.3)	305 (66.0)	171 (64.9)	196 (48.5)	231 (46.4)		Chlordane (peak 3)
373 (100.0)	375 (91.8)	371 (44.8)	377 (44.4)	237 (28.9)	272 (26.3)	239 (24.1)	65 (23.7)		Chlordane (peak 4)
373 (100.0)	375 (83.9)	377 (43.5)	371 (40.5)	66 (39.3)	237 (29.8)	272 (25.6)	65 (25.6)		Chlordane (peak 5)

222 (100.0)	152 (78.7)	224 (62.9)	75 (20.9)	151 (17.6)	93 (15.5)	150 (13.2)	223 (13.1)		PCB-1016 (peak 1)
186 (100.0)	258 (72.6)	256 (69.3)	221 (36.1)	150 (32.7)	188 (30.9)	75 (29.1)	222 (26.9)		PCB-1016 (peak 2)
186 (100.0)	258 (87.4)	256 (78.4)	221 (38.0)	150 (33.1)	75 (30.8)	260 (24.9)	151 (23.9)		PCB-1016 (peak 3)
258 (100.0)	256 (94.1)	186 (73.6)	260 (28.6)	150 (24.3)	188 (23.7)	75 (22.6)	93 (17.9)		PCB-1016 (peak 4)
258 (100.0)	256 (92.9)	186 (92.9)	257 (66.1)	92 (52.7)	110 (38.8)	260 (36.2)	150 (33.0)		PCB-1016 (peak 5)
154 (100.0)	153 (40.8)	152 (26.9)	155 (14.5)	76 (11.6)	151 (7.5)	51 (5.5)	77 (5.4)		PCB-1221 (peak 1)
188 (100.0)	152 (45.9)	190 (32.8)	153 (25.0)	76 (16.5)	189 (13.3)	151 (13.3)	75 (8.3)		PCB-1221 (peak 2)
188 (100.0)	152 (41.8)	190 (32.5)	153 (20.6)	76 (14.7)	189 (13.2)	151 (11.2)	63 (6.8)		PCB-1221 (peak 3)
152 (100.0)	222 (86.1)	224 (54.3)	187 (48.4)	75 (20.8)	151 (20.0)	93 (15.3)	150 (14.9)		PCB-1221 (peak 4)
222 (100.0)	152 (67.2)	224 (62.1)	75 (16.9)	151 (14.8)	223 (12.8)	150 (11.9)	226 (10.3)		PCB-1221 (peak 5)
188 (100.0)	152 (44.8)	190 (32.5)	153 (23.0)	76 (15.6)	151 (12.8)	189 (12.2)	63 (7.8)		PCB-1232 (peak 1)
188 (100.0)	152 (41.4)	190 (32.7)	153 (20.6)	76 (13.7)	189 (12.5)	151 (11.7)	63 (6.8)		PCB-1232 (peak 2)
222 (100.0)	152 (63.4)	224 (60.9)	75 (16.4)	151 (13.9)	223 (13.1)	150 (10.9)	93 (10.1)		PCB-1232 (peak 3)
186 (100.0)	222 (91.2)	258 (79.0)	256 (78.9)	152 (54.9)	224 (53.5)	223 (38.7)	150 (35.7)		PCB-1232 (peak 4)

Molecular Weight Index—continued

Partial spectrum								*M*	Substance
258 (100.0)	256 (91.1)	186 (60.1)	260 (27.7)	188 (21.0)	150 (20.3)	75 (19.3)	151 (14.4)		PCB-1232 (peak 5)
222 (100.0)	152 (65.0)	224 (62.5)	75 (16.6)	151 (16.0)	223 (12.4)	150 (11.4)	93 (11.3)		PCB-1242 (peak 1)
186 (100.0)	258 (84.4)	256 (80.7)	221 (37.6)	222 (35.1)	188 (34.2)	150 (31.4)	75 (29.7)		PCB-1242 (peak 2)
258 (100.0)	256 (92.9)	186 (63.9)	260 (28.0)	150 (21.7)	188 (20.8)	75 (20.8)	151 (15.7)		PCB-1242 (peak 3)
256 (100.0)	258 (89.0)	186 (81.9)	188 (33.5)	260 (29.1)	150 (25.3)	75 (24.7)	290 (22.5)		PCB-1242 (peak 4)
292 (100.0)	290 (92.2)	220 (61.0)	294 (53.1)	222 (40.2)	110 (27.3)	150 (19.1)	111 (15.6)		PCB-1242 (peak 5)
258 (100.0)	256 (97.0)	186 (66.3)	260 (29.4)	150 (23.6)	188 (22.2)	75 (21.5)	151 (16.7)		PCB-1248 (peak 1)
292 (100.0)	220 (93.4)	290 (81.5)	222 (60.7)	294 (48.6)	110 (37.9)	150 (27.9)	255 (27.8)		PCB-1248 (peak 2)
220 (100.0)	292 (88.8)	290 (70.2)	222 (67.0)	255 (45.7)	110 (45.7)	257 (43.6)	294 (39.9)		PCB-1248 (peak 3)
292 (100.0)	290 (90.5)	220 (64.9)	294 (50.8)	222 (37.2)	110 (22.5)	150 (20.0)	111 (17.2)		PCB-1248 (peak 4)
292 (100.0)	290 (88.0)	220 (62.9)	294 (50.0)	222 (38.3)	110 (29.1)	150 (18.6)	111 (16.7)		PCB-1248 (peak 5)
292 (100.0)	290 (90.9)	220 (63.6)	294 (56.2)	222 (40.2)	110 (31.3)	326 (23.7)	184 (21.2)		PCB-1254 (peak 1)

326 (100.0)	328 (71.0)	254 (63.8)	324 (62.1)	256 (58.8)	127 (33.1)	128 (31.6)	184 (26.7)	PCB-1254 (peak 2)
326 (100.0)	328 (76.4)	324 (66.4)	254 (41.5)	256 (31.2)	330 (20.9)	128 (18.8)	184 (16.7)	PCB-1254 (peak 3)
326 (100.0)	328 (73.1)	324 (62.3)	254 (40.2)	256 (35.8)	127 (27.9)	128 (22.0)	109 (23.4)	PCB-1254 (peak 4)
360 (100.0)	326 (91.7)	362 (78.7)	328 (68.1)	324 (54.2)	145 (47.8)	288 (47.3)	254 (44.4)	PCB-1254 (peak 5)
360 (100.0)	362 (73.5)	290 (60.3)	288 (42.5)	145 (41.3)	358 (37.7)	146 (27.4)	364 (27.1)	PCB-1260 (peak 1)
360 (100.0)	362 (76.4)	290 (62.7)	288 (46.5)	358 (42.4)	145 (39.7)	144 (32.2)	364 (30.6)	PCB-1260 (peak 2)
360 (100.0)	362 (75.5)	290 (59.4)	288 (43.1)	358 (42.8)	145 (37.8)	364 (30.3)	144 (26.1)	PCB-1260 (peak 3)
394 (100.0)	396 (96.6)	324 (66.5)	162 (57.1)	398 (49.2)	322 (47.0)	392 (46.6)	326 (45.5)	PCB-1260 (peak 4)
394 (100.0)	396 (96.7)	324 (66.2)	162 (59.3)	398 (49.2)	322 (46.3)	392 (46.1)	326 (45.6)	PCB-1260 (peak 5)
159 (100.0)	125 (100.0)	197 (80.4)	51 (79.4)	75 (78.4)	161 (73.2)	209 (68.0)	233 (62.9)	Toxaphene (peak 1)
100 (100.0)	99 (46.5)	109 (41.9)	102 (39.5)	197 (34.9)	199 (32.6)	125 (30.2)	77 (27.9)	Toxaphene (peak 2)
159 (100.0)	161 (89.9)	125 (77.8)	85 (74.7)	75 (74.7)	195 (66.7)	83 (65.7)	197 (57.6)	Toxaphene (peak 3)
85 (100.0)	83 (95.7)	100 (82.9)	73 (71.4)	329 (58.6)	159 (57.1)	51 (55.7)	75 (52.9)	Toxaphene (peak 4)
245 (100.0)	243 (100.0)	83 (97.3)	85 (78.4)	207 (64.9)	159 (48.6)	73 (45.9)	111 (43.2)	Toxaphene (peak 5)

Base-Peak Index

Partial spectrum								M	Substance
41 (100.0)	77 (93.2)	39 (54.6)	79 (36.4)	27 (20.5)	49 (14.0)	38 (13.4)	37 (7.9)	156	2-Bromo-1-chloropropane (IS)
45 (100.0)	121 (69.1)	41 (64.1)	77 (26.4)	123 (24.0)	79 (18.3)	49 (11.1)	107 (10.7)	170	Bis(2-chloroisopropyl) ether
49 (100.0)	130 (38.4)	128 (31.1)	51 (28.4)	79 (26.2)	81 (25.6)	93 (22.4)	95 (16.2)	128 (31.1)	Bromochloromethane (IS)
49 (100.0)	84 (60.1)	86 (38.3)	51 (28.4)	47 (22.0)	35 (14.6)	48 (9.6)	28 (6.6)	84 (60.1)	Methylene chloride
50 (100.0)	52 (34.2)	49 (13.3)	47 (11.9)	35 (8.4)	48 (5.0)	51 (4.6)	37 (3.0)	50 (100.0)	Chloromethane
53 (100.0)	52 (79.6)	26 (78.5)	51 (33.5)	27 (16.0)	25 (10.2)	38 (9.8)	50 (9.1)	53 (100.0)	Acrylonitrile
55 (100.0)	41 (38.2)	27 (31.2)	54 (28.6)	90 (26.5)	39 (20.2)	62 (19.5)	49 (17.0)	126	1,4-Dichlorobutane (IS)
56 (100.0)	27 (66.2)	55 (64.9)	26 (44.5)	29 (37.7)	37 (11.3)	38 (9.3)	53 (8.6)	56 (100.0)	Acrolein
61 (100.0)	96 (44.3)	63 (31.9)	98 (28.9)	60 (18.9)	26 (13.4)	35 (9.2)	62 (8.4)	96 (44.3)	1,1-Dichloroethylene
61 (100.0)	96 (60.2)	98 (39.3)	63 (32.3)	60 (29.7)	26 (16.3)	62 (12.2)	25 (8.6)	96 (60.2)	*trans*-1,2-Dichloroethylene
62 (100.0)	27 (39.8)	64 (31.1)	49 (24.9)	26 (13.1)	63 (12.5)	61 (8.8)	51 (7.8)	98 (7.3)	1,2-Dichloroethane
62 (100.0)	27 (76.4)	64 (30.8)	26 (24.2)	25 (10.3)	61 (8.4)	35 (7.3)	60 (6.5)	62 (100.0)	Vinyl chloride

63 (100.0)	43 (98.1)	44 (71.5)	65 (32.3)	42 (25.0)	62 (24.7)	106 (24.5)	57 (20.2)	106 (24.5)	2-Chloroethyl vinyl ether
63 (100.0)	27 (42.8)	65 (31.4)	26 (13.2)	83 (12.9)	61 (8.9)	62 (8.4)	85 (8.3)	98 (6.8)	1,1-Dichloroethane
63 (100.0)	62 (79.2)	41 (70.1)	76 (61.7)	39 (54.9)	27 (44.7)	65 (30.8)	64 (27.0)	112 (5.8)	1,2-Dichloropropane
64 (100.0)	28 (84.9)	27 (82.6)	29 (78.1)	26 (36.9)	66 (35.0)	49 (27.0)	51 (8.8)	64 (100.0)	Chloroethane
65 (100.0)	139 (83.6)	109 (43.4)	53 (29.7)	81 (27.6)	63 (27.3)	93 (18.9)	62 (14.8)	139 (83.6)	4-Nitrophenol
66 (100.0)	91 (42.7)	79 (38.9)	65 (31.9)	101 (29.6)	263 (29.4)	265 (22.3)	261 (22.3)	362	Aldrin
67 (100.0)	83 (55.6)	317 (50.8)	85 (45.7)	250 (36.1)	345 (35.6)	319 (33.7)	315 (33.3)	378 (2.8)	Endrin
67 (100.0)	250 (40.0)	345 (30.0)	197 (30.0)	135 (30.0)	95 (29.0)	248 (27.1)	66 (26.7)	378	Endrin aldehyde
70 (100.0)	43 (70.7)	130 (44.8)	42 (41.4)	41 (33.9)	101 (16.6)	58 (15.8)	113 (10.8)	130 (44.8)	*N*-Nitrosodi-n-propylamine
74 (100.0)	42 (54.1)	43 (24.9)	40 (3.8)	41 (3.1)	75 (3.0)	44 (2.8)	–	74 (100.0)	*N*-Nitrosodimethylamine
75 (100.0)	39 (50.3)	77 (28.9)	49 (21.2)	110 (19.6)	38 (13.9)	112 (12.2)	37 (12.1)	110 (19.6)	*cis*-1,3-Dichloropropene
75 (100.0)	39 (55.0)	77 (29.1)	49 (20.4)	110 (19.8)	38 (14.4)	37 (12.8)	112 (11.7)	110 (19.8)	*trans*-1,3-Dichloropropene
77 (100.0)	51 (37.6)	182 (24.2)	105 (15.5)	78 (11.3)	50 (9.5)	152 (5.7)	153 (3.4)	184	1,2-Diphenylhydrazine [azobenzene]

Base-Peak Index—continued

Partial spectrum								M	Substance
77 (100.0)	123 (75.6)	51 (43.8)	50 (16.0)	93 (14.8)	65 (14.0)	74 (7.9)	78 (6.7)	123 (75.6)	Nitrobenzene
77 (100.0)	51 (23.7)	182 (21.9)	105 (18.4)	78 (7.3)	152 (5.3)	50 (4.6)	183 (2.9)	198	*N*-Nitrosodiphenylamine [diphenylamine]
78 (100.0)	77 (25.9)	50 (20.8)	51 (20.6)	52 (18.8)	39 (9.3)	79 (6.4)	74 (5.8)	78 (100.0)	Benzene
79 (100.0)	49 (37.9)	81 (33.1)	51 (11.9)	29 (9.5)	78 (5.5)	80 (3.8)	48 (2.9)	114	Bis(chloromethyl) ether
79 (100.0)	82 (34.8)	81 (34.7)	77 (22.6)	108 (15.4)	53 (13.6)	107 (11.5)	263 (10.1)	378	Dieldrin
81 (100.0)	353 (38.6)	355 (35.9)	53 (25.0)	351 (24.6)	51 (20.7)	357 (16.2)	61 (15.7)	386 (2.3)	Heptachlor expoxide
82 (100.0)	54 (14.8)	138 (12.2)	41 (7.7)	53 (6.9)	83 (5.3)	67 (5.3)	55 (5.3)	138 (12.2)	Isophorone
83 (100.0)	85 (66.5)	47 (20.8)	48 (12.8)	79 (12.4)	81 (11.9)	129 (10.1)	87 (9.4)	162	Bromodichloromethane
83 (100.0)	85 (64.7)	47 (29.7)	35 (14.4)	48 (11.9)	49 (10.5)	87 (9.5)	37 (4.8)	118	Chloroform
83 (100.0)	85 (66.3)	95 (15.6)	60 (14.1)	61 (11.9)	96 (11.1)	168 (9.9)	87 (9.8)	166 (7.7)	1,1,2,2-Tetrachloroethane
85 (100.0)	87 (30.9)	50 (24.6)	35 (22.0)	31 (16.3)	47 (8.7)	101 (7.6)	66 (6.8)	120	Dichlorodifluoromethane
85 (100.0)	83 (95.7)	100 (82.9)	73 (71.4)	329 (58.6)	159 (57.1)	51 (55.7)	75 (52.9)		Toxaphene (peak 4)

91 (100.0)	106 (27.3)	65 (10.9)	51 (10.7)	77 (9.2)	78 (8.0)	92 (7.6)	39 (6.5)	106 (27.3)	Ethylbenzene
91 (100.0)	92 (57.8)	65 (14.4)	39 (12.8)	63 (10.4)	51 (8.3)	50 (5.9)	62 (5.0)	92 (57.8)	Toluene
93 (100.0)	63 (68.7)	95 (30.4)	123 (24.9)	65 (22.1)	125 (7.2)	106 (4.9)	49 (4.5)	172	Bis(2-chloroethoxy)methane
93 (100.0)	63 (57.8)	95 (34.4)	65 (16.6)	49 (5.4)	94 (3.8)	106 (3.0)	142 (2.9)	142 (2.9)	Bis(2-chloroethyl) ether
94 (100.0)	96 (89.3)	93 (21.6)	79 (21.3)	81 (20.5)	95 (14.1)	91 (9.0)	92 (4.3)	94 (100.0)	Bromomethane
94 (100.0)	66 (32.3)	65 (22.0)	95 (6.6)	63 (5.6)	55 (4.7)	50 (3.9)	51 (3.5)	94 (100.0)	Phenol
97 (100.0)	83 (76.0)	99 (61.7)	61 (54.7)	85 (53.0)	96 (28.3)	98 (20.6)	63 (17.7)	132 (15.7)	1,1,2-Trichloroethane
97 (100.0)	99 (60.9)	61 (51.2)	63 (16.3)	26 (11.8)	117 (10.7)	119 (10.2)	101 (10.2)	132	1,1,1-Trichloroethane
100 (100.0)	102 (35.2)	238 (33.0)	65 (30.4)	101 (23.7)	236 (19.3)	240 (17.4)	66 (14.4)		Chlordane (peak 1)
100 (100.0)	272 (54.9)	274 (44.7)	102 (38.1)	65 (35.0)	270 (33.2)	237 (20.4)	276 (17.3)		Chlordane (peak 2)
100 (100.0)	99 (46.5)	109 (41.9)	102 (39.5)	197 (34.9)	199 (32.6)	125 (30.2)	77 (27.9)		Toxaphene (peak 2)
100 (100.0)	272 (52.8)	65 (47.8)	274 (45.0)	102 (34.1)	270 (28.5)	237 (21.6)	135 (17.7)	370 (2.2)	Heptachlor
101 (100.0)	103 (59.1)	66 (17.6)	35 (12.8)	105 (11.0)	47 (10.9)	31 (9.3)	68 (6.0)	136	Trichlorofluoromethane
107 (100.0)	142 (76.4)	77 (43.7)	144 (24.8)	51 (17.4)	78 (12.1)	79 (11.8)	143 (9.0)	142 (76.4)	*p*-Chloro-*m*-cresol

Base-Peak Index—continued

Partial spectrum								M	Substance
107 (100.0)	122 (92.1)	121 (49.6)	77 (26.9)	91 (19.4)	79 (15.0)	78 (9.0)	51 (8.9)	122 (92.1)	2,4-Dimethylphenol
109 (100.0)	181 (89.3)	183 (82.7)	111 (82.4)	217 (65.5)	51 (48.8)	85 (46.0)	83 (38.6)	288	β-BHC
112 (100.0)	77 (67.6)	114 (31.4)	51 (25.8)	50 (25.0)	45 (11.9)	74 (11.4)	75 (9.7)	112 (100.0)	Chlorobenzene
117 (100.0)	119 (93.7)	121 (30.7)	82 (28.3)	47 (25.6)	35 (18.7)	84 (18.3)	49 (8.9)	152	Carbon tetrachloride
117 (100.0)	119 (83.0)	201 (57.8)	203 (37.5)	199 (34.7)	166 (32.3)	94 (30.5)	47 (30.4)	234	Hexachloroethane
128 (100.0)	130 (31.2)	64 (29.3)	63 (16.1)	92 (12.3)	129 (6.7)	73 (5.0)	65 (4.8)	128 (100.0)	2-Chlorophenol
128 (100.0)	127 (14.0)	129 (11.4)	102 (8.9)	126 (7.8)	51 (5.8)	75 (4.9)	63 (4.7)	128 (100.0)	Naphthalene
129 (100.0)	127 (76.2)	97 (39.6)	83 (36.0)	61 (35.2)	79 (24.1)	131 (23.7)	81 (23.4)	206 (2.6)	Dibromochloromethane
130 (100.0)	132 (98.5)	95 (84.3)	97 (55.1)	60 (37.4)	134 (30.2)	62 (13.1)	35 (10.2)	130 (100.0)	Trichloroethylene
139 (100.0)	65 (68.0)	63 (59.2)	81 (45.2)	53 (34.8)	64 (34.2)	109 (26.7)	51 (19.0)	139 (100.0)	2-Nitrophenol
146 (100.0)	148 (60.1)	75 (52.5)	111 (51.5)	50 (37.5)	74 (31.8)	113 (18.0)	73 (15.9)	146 (100.0)	1,2-Dichlorobenzene
146 (100.0)	148 (58.3)	75 (50.6)	111 (49.7)	50 (36.5)	74 (32.4)	73 (18.1)	113 (16.0)	146 (100.0)	1,3-Dichlorobenzene

146 (100.0)	148 (58.2)	111 (33.4)	75 (20.8)	74 (13.0)	113 (11.0)	150 (10.1)	50 (9.4)	146 (100.0)	1,4-Dichlorobenzene
149 (100.0)	91 (69.5)	206 (24.4)	65 (16.4)	123 (14.9)	104 (14.2)	132 (12.7)	150 (11.2)	312 (2.0)	Butyl benzyl phthalate
149 (100.0)	167 (43.1)	57 (23.3)	71 (15.4)	70 (15.0)	43 (14.3)	279 (13.7)	41 (13.6)	390	Bis(2-Ethylhexyl) phthalate
149 (100.0)	150 (9.0)	41 (6.3)	223 (5.7)	104 (5.0)	76 (4.8)	205 (4.5)	65 (3.4)	278	Di-n-butyl phthalate
149 (100.0)	177 (23.7)	65 (10.6)	176 (10.5)	150 (10.2)	76 (9.6)	105 (8.5)	121 (7.0)	222 (4.6)	Diethyl phthalate
149 (100.0)	279 (10.7)	150 (9.7)	43 (7.6)	41 (7.5)	57 (5.5)	71 (3.8)	69 (3.2)	390	Di-n-octyl phthalate
152 (100.0)	151 (17.5)	150 (13.5)	153 (12.6)	76 (9.3)	75 (3.8)	74 (3.7)	126 (3.6)	152 (100.0)	Acenaphthylene
152 (100.0)	222 (86.1)	224 (54.3)	187 (48.4)	75 (20.8)	151 (20.0)	93 (15.3)	150 (14.9)		PCB-1221 (peak 4)
153 (100.0)	154 (96.5)	152 (44.6)	76 (19.0)	151 (16.7)	155 (14.0)	63 (8.9)	150 (8.5)	154 (96.5)	Acenaphthene
154 (100.0)	153 (40.8)	152 (26.9)	155 (14.5)	76 (11.6)	151 (7.5)	51 (5.5)	77 (5.4)		PCB-1221 (peak 1)
159 (100.0)	125 (100.0)	197 (80.4)	51 (79.4)	75 (78.4)	161 (73.2)	209 (68.0)	233 (62.9)		Toxaphene (peak 1)
159 (100.0)	161 (89.9)	125 (77.8)	85 (74.7)	75 (74.7)	195 (66.7)	83 (65.7)	197 (57.6)		Toxaphene (peak 3)
162 (100.0)	127 (32.1)	164 (29.0)	126 (17.3)	163 (10.9)	63 (8.3)	75 (7.6)	81 (6.7)	162 (100.0)	2-Chloronaphthalene
162 (100.0)	164 (62.2)	63 (45.4)	98 (31.5)	126 (13.9)	99 (12.3)	62 (12.2)	73 (11.0)	162 (100.0)	2,4-Dichlorophenol

Base-Peak Index—continued

Partial spectrum								*M*	Substance
163 (100.0)	77 (17.7)	194 (10.2)	164 (9.5)	92 (8.0)	76 (7.6)	50 (7.2)	133 (6.2)	194 (10.2)	Dimethyl phthalate
165 (100.0)	89 (64.4)	63 (53.0)	90 (22.7)	119 (18.5)	78 (18.4)	51 (18.1)	64 (17.2)	182 (9.1)	2,4-Dinitrotoluene
165 (100.0)	63 (47.7)	89 (41.7)	90 (30.4)	77 (23.9)	51 (22.2)	64 (21.0)	78 (20.8)	182 (2.5)	2,6-Dinitrotoluene
166 (100.0)	129 (77.1)	164 (74.2)	131 (73.4)	168 (54.3)	94 (53.8)	47 (35.7)	96 (33.6)	164 (74.2)	1,1,2,2-Tetrachloroethene
166 (100.0)	165 (90.9)	163 (14.2)	164 (12.6)	167 (12.4)	139 (6.5)	115 (4.2)	63 (3.9)	166 (100.0)	Fluorene
173 (100.0)	171 (51.9)	175 (50.7)	79 (30.6)	91 (29.7)	93 (29.6)	81 (28.9)	94 (16.9)	250 (3.3)	Bromoform
178 (100.0)	176 (17.7)	179 (13.9)	177 (8.6)	76 (8.2)	89 (7.6)	152 (7.2)	151 (5.6)	178 (100.0)	Anthracene
178 (100.0)	176 (17.9)	179 (14.2)	76 (10.3)	177 (10.1)	152 (9.4)	89 (8.8)	88 (7.3)	178 (100.0)	Phenanthrene
180 (100.0)	182 (96.3)	184 (30.2)	145 (29.0)	109 (22.1)	74 (21.6)	147 (17.1)	75 (12.8)	180 (100.0)	1,2,4-Trichlorobenzene
181 (100.0)	183 (91.7)	219 (84.7)	217 (66.0)	109 (65.3)	111 (62.4)	51 (43.0)	85 (36.9)	288	α-BHC
181 (100.0)	183 (98.5)	111 (96.9)	219 (83.0)	109 (74.4)	217 (58.3)	51 (55.6)	73 (55.4)	288 (2.9)	γ-BHC
181 (100.0)	183 (94.0)	219 (89.6)	109 (85.0)	111 (74.9)	217 (64.5)	51 (55.2)	85 (49.7)	288	δ-BHC

184 (100.0)	185 (12.3)	183 (11.5)	92 (6.8)	156 (6.4)	91 (5.9)	167 (5.1)	166 (4.9)	184 (100.0)	Benzidine
184 (100.0)	63 (67.9)	53 (61.2)	91 (39.3)	62 (35.7)	107 (33.9)	154 (31.3)	79 (31.3)	184 (100.0)	2,4-Dinitrophenol
186 (100.0)	258 (72.6)	256 (69.3)	221 (36.1)	150 (32.7)	188 (30.9)	75 (29.1)	222 (26.9)		PCB-1016 (peak 2)
186 (100.0)	258 (87.4)	256 (78.4)	221 (38.0)	150 (33.1)	75 (30.8)	260 (24.9)	151 (23.9)		PCB-1016 (peak 3)
186 (100.0)	222 (91.2)	258 (79.0)	256 (78.9)	152 (54.9)	224 (53.5)	223 (38.7)	150 (35.7)		PCB-1232 (peak 4)
186 (100.0)	258 (84.4)	256 (80.7)	221 (37.6)	222 (35.1)	188 (34.2)	150 (31.4)	75 (29.7)		PCB-1242 (peak 2)
188 (100.0)	187 (19.1)	80 (16.1)	184 (15.7)	189 (15.0)	94 (14.9)	160 (11.8)	66 (11.5)	188 (100.0)	Anthracene-d_{10} (IS)
188 (100.0)	152 (45.9)	190 (32.8)	153 (25.0)	76 (16.5)	189 (13.3)	151 (13.3)	75 (8.3)		PCB-1221 (peak 2)
188 (100.0)	162 (41.8)	190 (32.5)	153 (20.6)	76 (14.7)	189 (13.2)	151 (11.2)	63 (6.8)		PCB-1221 (peak 3)
188 (100.0)	152 (44.8)	190 (32.5)	153 (23.0)	76 (15.6)	151 (12.8)	189 (12.2)	63 (7.8)		PCB-1232 (peak 1)
188 (100.0)	152 (41.4)	190 (32.7)	153 (20.6)	76 (13.7)	189 (12.5)	151 (11.7)	63 (6.8)		PCB-1232 (peak 2)
195 (100.0)	170 (79.3)	207 (75.2)	239 (71.0)	237 (69.0)	159 (67.6)	241 (66.9)	197 (62.8)	404 (2.1)	α-Endosulfan
195 (100.0)	197 (80.5)	159 (75.6)	160 (65.9)	207 (63.4)	170 (53.7)	239 (51.2)	237 (51.2)	404 (4.9)	β-Endosulfan
196 (100.0)	198 (97.5)	97 (56.8)	132 (42.7)	200 (33.9)	134 (28.0)	62 (25.7)	99 (22.9)	196 (100.0)	2,4,6-Trichlorophenol

Base-Peak Index—continued

Partial spectrum								*M*	Substance
198 (100.0)	51 (52.3)	105 (44.5)	53 (41.9)	121 (41.5)	77 (24.7)	52 (24.2)	50 (100.0)	198 (100.0)	4,6-Dinitro-*o*-cresol
202 (100.0)	200 (20.2)	203 (18.8)	101 (15.9)	201 (13.9)	100 (11.3)	88 (8.7)	87 (5.2)	202 (100.0)	Fluoranthene
202 (100.0)	200 (22.2)	201 (19.7)	203 (19.5)	101 (15.4)	100 (13.0)	88 (5.5)	199 (4.5)	202 (100.0)	Pyrene
204 (100.0)	141 (50.0)	206 (36.9)	77 (28.7)	51 (20.3)	115 (12.7)	75 (12.7)	205 (12.0)	204 (100.0)	4-Chlorophenyl phenyl ether
220 (100.0)	292 (88.8)	290 (70.2)	222 (67.0)	255 (45.7)	110 (45.7)	257 (43.6)	294 (39.9)		PCB-1248 (peak 3)
222 (100.0)	152 (78.7)	224 (62.9)	75 (20.9)	151 (17.6)	93 (15.5)	150 (13.2)	223 (13.1)		PCB-1016 (peak 1)
222 (100.0)	152 (67.2)	224 (62.1)	75 (16.9)	151 (14.8)	223 (12.8)	150 (11.9)	226 (10.3)		PCB-1221 (peak 5)
222 (100.0)	152 (63.4)	224 (60.9)	75 (16.4)	151 (13.9)	223 (13.1)	150 (10.9)	93 (10.1)		PCB-1232 (peak 3)
222 (100.0)	152 (65.0)	224 (62.5)	75 (16.6)	151 (16.0)	223 (12.4)	150 (11.4)	93 (11.3)		PCB-1242 (peak 1)
225 (100.0)	223 (67.2)	227 (64.1)	190 (45.9)	188 (37.9)	118 (34.8)	260 (33.6)	262 (23.6)	258 (20.8)	Hexachlorobutadiene
228 (100.0)	226 (26.5)	113 (23.0)	114 (22.8)	229 (20.6)	101 (12.5)	112 (10.0)	100 (9.5)	228 (100.0)	Benzo[*a*]anthracene
228 (100.0)	226 (30.9)	113 (26.3)	114 (22.2)	229 (20.3)	227 (12.5)	112 (11.0)	100 (10.3)	228 (100.0)	Chrysene

232 (100.0)	303 (94.8)	230 (94.8)	169 (75.3)	305 (66.0)	171 (64.9)	196 (48.5)	231 (46.4)		Chlordane (peak 3)
235 (100.0)	165 (58.4)	237 (57.0)	75 (21.2)	199 (13.5)	88 (12.3)	236 (12.2)	82 (10.2)	318 (2.6)	4,4′-DDD
235 (100.0)	237 (56.6)	165 (53.8)	75 (17.6)	176 (13.3)	236 (12.5)	199 (11.9)	212 (11.0)	352	4,4′-DDT
237 (100.0)	235 (73.0)	239 (69.7)	95 (28.4)	130 (22.6)	241 (22.2)	272 (17.8)	60 (16.9)	270 (9.7)	Hexachlorocyclopentadiene
245 (100.0)	243 (100.0)	83 (97.3)	85 (78.4)	207 (64.9)	159 (48.6)	73 (45.9)	111 (43.2)		Toxaphene (peak 5)
246 (100.0)	318 (76.8)	248 (72.9)	316 (57.1)	176 (57.0)	105 (40.4)	320 (35.0)	75 (23.8)	316 (57.1)	4,4′-DDE
248 (100.0)	250 (88.6)	141 (69.5)	77 (58.5)	51 (37.3)	115 (29.3)	50 (24.4)	63 (22.5)	248 (100.0)	4-Bromophenyl phenyl ether
252 (100.0)	126 (23.0)	250 (21.7)	253 (20.5)	125 (14.7)	113 (10.8)	124 (9.7)	112 (8.8)	252 (100.0)	Benzo[*b*]fluoranthene
252 (100.0)	250 (20.9)	253 (20.7)	126 (14.9)	125 (10.2)	113 (7.1)	112 (6.5)	251 (6.2)	252 (100.0)	Benzo[*k*]fluoranthene
252 (100.0)	250 (22.1)	253 (22.0)	126 (16.9)	125 (10.4)	113 (9.7)	251 (8.2)	112 (6.8)	252 (100.0)	Benzo[*a*]pyrene
252 (100.0)	254 (59.1)	253 (17.8)	126 (17.3)	127 (15.5)	154 (13.7)	91 (12.6)	181 (10.8)	252 (100.0)	3,3-Dichlorobenzidine
256 (100.0)	258 (89.0)	186 (81.9)	188 (33.5)	260 (29.1)	150 (25.3)	75 (24.7)	290 (22.5)		PCB-1242 (peak 4)
258 (100.0)	256 (94.1)	186 (73.6)	260 (28.6)	150 (24.3)	188 (23.7)	75 (22.6)	93 (17.9)		PCB-1016 (peak 4)
258 (100.0)	256 (92.9)	186 (92.9)	257 (66.1)	92 (52.7)	110 (38.8)	260 (36.2)	150 (33.0)		PCB-1016 (peak 5)

Base-Peak Index—continued

Partial spectrum								M	Substance
258 (100.0)	256 (91.1)	186 (60.1)	260 (27.7)	188 (21.0)	150 (20.3)	75 (19.3)	151 (14.4)		PCB-1232 (peak 5)
258 (100.0)	256 (92.9)	186 (63.9)	260 (28.0)	150 (21.7)	188 (20.8)	75 (20.8)	151 (15.7)		PCB-1242 (peak 3)
258 (100.0)	256 (97.0)	186 (66.3)	260 (29.4)	150 (23.6)	188 (22.2)	75 (21.5)	151 (16.7)		PCB-1248 (peak 1)
266 (100.0)	268 (74.3)	264 (70.0)	165 (45.5)	95 (39.5)	167 (38.7)	60 (26.5)	270 (26.1)	264 (70.0)	Pentachlorophenol
272 (100.0)	274 (97.7)	229 (76.1)	270 (62.5)	227 (62.5)	237 (56.8)	387 (48.9)	239 (48.9)	420 (5.7)	Endosulfan sulfate
276 (100.0)	277 (23.7)	274 (21.3)	138 (19.4)	137 (17.8)	275 (11.4)	136 (8.9)	272 (5.2)	276 (100.0)	Benzo[*ghi*]perylene
276 (100.0)	277 (24.1)	274 (21.5)	138 (18.2)	137 (13.9)	275 (9.2)	136 (7.8)	125 (5.2)	276 (100.0)	Indeno[1,2,3-*cd*]pyrene
278 (100.0)	139 (38.6)	138 (27.2)	279 (23.4)	276 (21.9)	137 (16.3)	125 (15.2)	113 (8.8)	278 (100.0)	Dibenzo[*a*,*h*]anthracene
284 (100.0)	286 (83.7)	282 (50.7)	142 (35.3)	288 (32.8)	249 (28.1)	107 (25.6)	144 (22.6)	282 (50.7)	Hexachlorobenzene
292 (100.0)	290 (92.2)	220 (61.0)	294 (53.1)	222 (40.2)	110 (27.3)	150 (19.1)	111 (15.6)		PCB-1242 (peak 5)
292 (100.0)	220 (93.4)	290 (81.5)	222 (60.7)	294 (48.6)	110 (37.9)	150 (27.9)	255 (27.8)		PCB-1248 (peak 2)
292 (100.0)	290 (90.5)	220 (64.9)	294 (50.8)	222 (37.2)	110 (22.5)	150 (20.0)	111 (17.2)		PCB-1248 (peak 4)

292 (100.0)	290 (92.2)	220 (61.0)	294 (53.1)	222 (38.3)	110 (29.1)	150 (18.6)	111 (16.7)		PCB-1248 (peak 5)
292 (100.0)	290 (90.9)	220 (63.6)	294 (56.2)	222 (40.2)	110 (31.3)	326 (23.7)	184 (21.2)		PCB-1254 (peak 1)
322 (100.0)	320 (78.5)	324 (47.9)	74 (37.4)	97 (36.0)	259 (34.3)	257 (32.9)	194 (31.6)	320 (78.5)	2,3,7,8-Tetrachlorodibenzo-*p*-dioxin
326 (100.0)	328 (71.0)	254 (63.8)	324 (62.1)	256 (58.8)	127 (33.1)	128 (31.6)	184 (26.7)		PCB-1254 (peak 2)
326 (100.0)	328 (76.4)	324 (66.4)	254 (41.5)	256 (31.2)	330 (20.9)	128 (18.8)	184 (16.7)		PCB-1254 (peak 3)
326 (100.0)	328 (73.1)	324 (62.3)	254 (40.2)	256 (35.8)	127 (27.9)	128 (22.0)	109 (23.4)		PCB-1254 (peak 4)
360 (100.0)	326 (91.7)	362 (78.7)	328 (68.1)	324 (54.2)	145 (47.8)	288 (47.3)	254 (44.4)		PCB-1254 (peak 5)
360 (100.0)	362 (73.5)	290 (60.3)	288 (42.5)	145 (41.3)	358 (37.7)	146 (27.4)	364 (27.1)		PCB-1260 (peak 1)
360 (100.0)	362 (76.4)	290 (62.7)	288 (46.5)	358 (42.4)	145 (39.7)	144 (32.2)	364 (30.6)		PCB-1260 (peak 2)
360 (100.0)	362 (75.5)	290 (59.4)	288 (43.1)	358 (42.8)	145 (37.8)	364 (30.3)	144 (26.1)		PCB-1260 (peak 3)
373 (100.0)	375 (91.8)	371 (44.8)	377 (44.4)	237 (28.9)	272 (26.3)	239 (24.1)	65 (23.7)		Chlordane (peak 4)
373 (100.0)	375 (83.9)	377 (43.5)	371 (40.5)	66 (39.3)	237 (29.8)	272 (25.6)	65 (25.6)		Chlordane (peak 5)
394 (100.0)	396 (96.6)	324 (66.5)	162 (57.1)	398 (49.2)	322 (47.0)	392 (46.6)	326 (45.5)		PCB-1260 (peak 4)
394 (100.0)	396 (96.7)	324 (66.2)	162 (59.3)	398 (49.2)	322 (46.3)	392 (46.1)	326 (45.6)		PCB-1260 (peak 5)

Second-Peak Index

Partial spectrum								*M*	Substance
56 (100.0)	**27** (66.2)	55 (64.9)	26 (44.5)	29 (37.7)	37 (11.3)	38 (9.3)	53 (8.6)	56 (100.0)	Acrolein
63 (100.0)	**27** (42.8)	65 (31.4)	26 (13.2)	83 (12.9)	61 (8.9)	62 (8.4)	85 (8.3)	98 (6.8)	1,1-Dichloroethane
62 (100.0)	**27** (39.8)	64 (31.1)	49 (24.9)	26 (13.1)	63 (12.5)	61 (8.8)	51 (7.8)	98 (7.3)	1,2-Dichloroethane
62 (100.0)	**27** (76.4)	64 (30.8)	26 (24.2)	25 (10.3)	61 (8.4)	35 (7.3)	60 (6.5)	62 (100.0)	Vinyl chloride
64 (100.0)	**28** (84.9)	27 (82.6)	29 (78.1)	26 (36.9)	66 (35.0)	49 (27.0)	51 (8.8)	64 (100.0)	Chloroethane
75 (100.0)	**39** (50.3)	77 (28.9)	49 (21.2)	110 (19.6)	38 (13.9)	112 (12.2)	37 (12.1)	110 (10.6)	*cis*-1,3-Dichloropropene
75 (100.0)	**39** (55.0)	77 (29.1)	49 (20.4)	110 (19.8)	38 (14.4)	37 (12.8)	112 (11.7)	110 (19.8)	*trans*-1,3-Dichloropropene
55 (100.0)	**41** (38.2)	27 (31.2)	54 (28.6)	90 (26.5)	39 (20.2)	62 (19.5)	49 (17.0)	126	1,4-Dichlorobutane (IS)
74 (100.0)	**42** (54.1)	43 (24.9)	40 (3.8)	41 (3.1)	75 (3.0)	44 (2.8)	–	74 (100.0)	*N*-Nitrosodimethylamine
63 (100.0)	**43** (98.1)	44 (71.5)	65 (32.3)	42 (25.0)	62 (24.7)	106 (24.5)	57 (20.2)	106 (24.5)	2-Chloroethyl vinyl ether
70 (100.0)	**43** (70.7)	130 (44.8)	42 (41.4)	41 (33.9)	101 (16.6)	58 (15.8)	113 (10.8)	130 (44.8)	*N*-Nitrosodi-n-propylamine
79 (100.0)	**49** (37.9)	81 (33.1)	51 (11.9)	29 (9.5)	78 (5.5)	80 (3.8)	48 (2.9)	114	Bis(chloromethyl) ether

198 (100.0)	**51 (52.3)**	105 (44.5)	53 (41.9)	121 (41.5)	77 (24.7)	52 (24.2)	50 (24.2)	198 (100.0)	4,6-Dinitro-*o*-cresol
77 (100.0)	**51 (37.6)**	182 (24.2)	105 (15.5)	78 (11.3)	50 (9.5)	152 (5.7)	153 (3.4)	184	1,2-Diphenylhydrazine [azobenzene]
77 (100.0)	**51 (23.7)**	182 (21.9)	105 (18.4)	78 (7.3)	152 (5.3)	50 (4.6)	183 (2.9)	198	*N*-Nitrosodiphenylamine [diphenylamine]
53 (100.0)	**52 (79.6)**	26 (78.5)	51 (33.5)	27 (16.0)	25 (10.2)	38 (9.8)	50 (9.1)	53 (100.0)	Acrylonitrile
50 (100.0)	**52 (34.2)**	49 (13.3)	47 (11.9)	35 (8.4)	48 (5.0)	51 (4.6)	37 (3.0)	50 (100.0)	Chloromethane
82 (100.0)	**54 (14.8)**	138 (12.2)	41 (7.7)	53 (6.9)	83 (5.3)	67 (5.3)	55 (5.3)	138 (12.2)	Isophorone
63 (100.0)	**62 (79.2)**	41 (70.1)	76 (61.7)	39 (54.9)	27 (44.7)	65 (30.8)	64 (27.0)	112 (5.8)	1,2-Dichloropropane
93 (100.0)	**63 (68.7)**	95 (30.4)	123 (24,9)	65 (22.1)	125 (7.2)	106 (4.9)	49 (4.5)	172	Bis(2-chloroethoxy)methane
93 (100.0)	**63 (57.8)**	95 (34.4)	65 (16.6)	49 (5.4)	94 (3.8)	106 (3.0)	142 (2.9)	142 (2.9)	Bis(2-chloroethyl) ether
184 (100.0)	**63 (67.9)**	53 (61.2)	91 (39.3)	62 (35.7)	107 (33.9)	154 (31.3)	79 (31.3)	184 (100.0)	2,4-Dinitrophenol
165 (100.0)	**63 (47.7)**	89 (41.7)	90 (30.4)	77 (23.9)	51 (22.2)	64 (21.0)	78 (20.8)	182 (2.5)	2,6-Dinitrotoluene
139 (100.0)	**65 (68.0)**	63 (59.2)	81 (45.2)	53 (34.8)	64 (34.2)	109 (26.7)	51 (19.0)	139 (100.0)	2-Nitrophenol
94 (100.0)	**66 (32.3)**	65 (22.0)	95 (6.6.)	63 (5.6)	55 (4.7)	50 (3.9)	51 (3.5)	94 (100.0)	Phenol

Second-Peak Index—continued

Partial spectrum								M	Substance
78 (100.0)	77 (25.9)	50 (20.8)	51 (20.6)	52 (18.8)	39 (9.3)	79 (6.4)	74 (5.8)	78 (100.0)	Benzene
41 (100.0)	77 (93.2)	39 (54.6)	79 (36.4)	27 (20.5)	49 (14.0)	38 (13.4)	37 (7.9)	156	2-Bromo-1-chloropropane (IS)
112 (100.0)	77 (67.6)	114 (31.4)	51 (25.8)	50 (25.0)	45 (11.9)	74 (11.4)	75 (9.7)	112 (100.0)	Chlorobenzene
163 (100.0)	77 (17.7)	194 (10.2)	164 (9.5)	92 (8.0)	76 (7.6)	50 (7.2)	133 (6.2)	194 (10.2)	Dimethyl phthalate
79 (100.0)	82 (34.8)	81 (34.7)	77 (22.6)	108 (15.4)	53 (13.6)	107 (11.5)	263 (10.1)	378	Dieldrin
67 (100.0)	83 (55.6)	317 (50.8)	85 (45.7)	250 (36.1)	345 (35.6)	319 (33.7)	315 (33.3)	378 (2.8)	Endrin
85 (100.0)	83 (95.7)	100 (82.9)	73 (71.4)	329 (58.6)	159 (57.1)	51 (55.7)	75 (52.9)		Toxaphene (peak 4)
97 (100.0)	83 (76.0)	99 (61.7)	61 (54.7)	85 (53.0)	96 (28.3)	98 (20.6)	63 (17.7)	132 (15.7)	1,1,2-Trichloroethane
49 (100.0)	84 (60.1)	86 (38.3)	51 (28.4)	47 (22.0)	35 (14.6)	48 (9.6)	28 (6.6)	84 (60.1)	Methylene chloride
83 (100.0)	85 (66.5)	47 (20.8)	48 (12.8)	79 (12.4)	81 (11.9)	129 (10.1)	87 (9.4)	162	Bromodichloromethane
83 (100.0)	85 (64.7)	47 (29.7)	35 (14.4)	48 (11.9)	49 (10.5)	87 (9.5)	37 (4.8)	118	Chloroform
83 (100.0)	85 (66.3)	95 (15.6)	60 (14.1)	61 (11.9)	96 (11.1)	168 (9.9)	87 (9.8)	166 (7.7)	1,1,2,2-Tetrachloroethane

85 (100.0)	**87** (30.9)	50 (24.6)	35 (22.0)	31 (16.3)	47 (8.7)	101 (7.6)	66 (6.8)	120	Dichlorodifluoromethane
165 (100.0)	**89** (64.4)	63 (53.0)	90 (22.7)	119 (18.5)	78 (18.4)	51 (18.1)	64 (17.2)	182 (9.1)	2,4-Dinitrotoluene
66 (100.0)	**91** (42.7)	79 (38.9)	65 (31.9)	101 (29.6)	263 (29.4)	265 (22.3)	261 (22.3)	362	Aldrin
149 (100.0)	**91** (69.5)	206 (24.4)	65 (16.4)	123 (14.9)	104 (14.2)	132 (12.7)	150 (11.2)	312 (2.0)	Buty benzyl phthalate
91 (100.0)	**92** (57.8)	65 (14.4)	39 (12.8)	63 (10.4)	51 (8.3)	50 (5.9)	62 (5.0)	92 (57.8)	Toluene
94 (100.00	**96** (89.3)	93 (21.6)	79 (21.3)	81 (20.5)	95 (14.1)	91 (9.0)	92 (4.3)	94 (100.0)	Bromomethane
61 (100.0)	**96** (44.3)	63 (31.9)	98 (28.9)	60 (18.9)	26 (13.4)	35 (9.2)	62 (8.4)	96 (44.3)	1,1-Dichloroethylene
61 (100.0)	**96** (60.2)	98 (39.3)	63 (32.3)	60 (29.7)	26 (16.3)	62 (12.2)	25 (8.6)	96 (60.2)	*trans*-1,2-Dichloroethylene
100 (100.0)	**99** (46.5)	109 (41.9)	102 (39.5)	197 (34.9)	199 (32.6)	125 (30.2)	77 (27.9)		Toxaphene (peak 2)
97 (100.0)	**99** (60.9)	61 (51.2)	63 (16.3)	26 (11.8)	117 (10.7)	119 (10.2)	101 (10.2)	132	1,1,1-Trichloroethane
100 (100.0)	**102** (35.2)	238 (33.0)	65 (30.4)	101 (23.7)	236 (19.3)	240 (17.4)	66 (14.4)		Chlordane (peak 1)
101 (100.0)	**103** (59.1)	66 (17.6)	35 (12.8)	105 (11.0)	47 (10.9)	31 (9.3)	68 (6.0)	136	Trichlorofluoromethane
91 (100.0)	**106** (27.3)	65 (10.9)	51 (10.7)	77 (9.2)	78 (8.0)	92 (7.6)	39 (6.5)	106 (27.3)	Ethylbenzene
117 (100.0)	**119** (93.7)	121 (30.7)	82 (28.3)	47 (25.6)	35 (18.7)	84 (18.3)	49 (8.9)	152	Carbon tetrachloride

Second-Peak Index—continued

Partial spectrum								M	Substance
117 (100.0)	**119 (83.0)**	201 (57.8)	203 (37.5)	199 (34.7)	166 (32.3)	94 (30.5)	47 (30.4)	234	Hexachloroethane
45 (100.0)	**121 (69.1)**	41 (64.1)	77 (26.4)	123 (24.0)	79 (18.3)	49 (11.1)	107 (10.7)	170	Bis(2-chloroisopropyl) ether
107 (100.0)	**122 (92.1)**	121 (49.6)	77 (26.9)	91 (19.4)	79 (15.0)	78 (9.0)	51 (8.9)	122 (92.1)	2,4-Dimethylphenol
77 (100.0)	**123 (75.6)**	51 (43.8)	50 (16.0)	93 (14.8)	65 (14.0)	74 (7.9)	78 (6.7)	123 (75.6)	Nitrobenzene
159 (100.0)	**125 (100.0)**	197 (80.4)	51 (79.4)	75 (78.4)	161 (73.2)	209 (68.0)	233 (62.9)		Toxaphene (peak 1)
252 (100.0)	**126 (23.0)**	250 (21.7)	253 (20.5)	125 (14.7)	113 (10.8)	124 (9.7)	112 (8.8)	252 (100.0)	Benzo[*b*]fluoroanthene
162 (100.0)	**127 (32.1)**	164 (29.0)	126 (17.3)	163 (10.9)	63 (8.3)	75 (7.6)	81 (6.7)	162 (100.0)	2-Chloronaphthalene
129 (100.0)	**127 (76.2)**	97 (39.6)	83 (36.0)	61 (35.2)	79 (24.1)	131 (23.7)	81 (23.4)	206 (2.6)	Dibromochloromethane
128 (100.0)	**127 (14.0)**	129 (11.4)	102 (8.9)	126 (7.8)	51 (5.8)	75 (4.9)	63 (4.7)	128 (100.0)	Naphthalene
166 (100.0)	**129 (77.1)**	164 (74.2)	131 (73.4)	168 (54.3)	94 (53.8)	47 (35.7)	96 (33.6)	164 (74.2)	1,1,2,2-Tetrachloroethene
49 (100.0)	**130 (38.4)**	128 (31.1)	51 (28.4)	79 (26.2)	81 (25.6)	93 (22.4)	95 (16.2)	128 (31.1)	Bromochloromethane (IS)
128 (100.0)	**130 (31.2)**	64 (29.3)	63 (16.1)	92 (12.3)	129 (6.7)	73 (5.0)	65 (4.8)	128 (100.0)	2-Chlorophenol

130 (100.0)	**132** (98.5)	95 (84.3)	97 (55.1)	60 (37.4)	134 (39.2)	62 (13.1)	35 (10.2)	130 (100.0)	Trichloroethylene
278 (100.0)	**139** (38.6)	138 (27.2)	279 (23.4)	276 (21.9)	137 (16.3)	125 (15.2)	113 (8.8)	278 (100.0)	Dibenzo[*a*,*h*]anthracene
65 (100.0)	**139** (83.6)	109 (43.4)	53 (29.7)	81 (27.6)	63 (27.3)	93 (18.9)	64 (14.8)	139 (83.6)	4-Nitrophenol
204 (100.0)	**141** (50.0)	206 (36.9)	77 (28.7)	51 (20.3)	115 (12.7)	75 (12.7)	205 (12.0)	204 (100.0)	4-Chlorophenyl phenyl ether
107 (100.0)	**142** (76.4)	77 (43.7)	144 (24.8)	51 (17.4)	78 (12.1)	79 (11.8)	143 (9.0)	142 (76.4)	*p*-Chloro-*m*-cresol
146 (100.0)	**148** (60.1)	75 (52.5)	111 (51.5)	50 (37.5)	74 (31.8)	113 (18.0)	73 (15.9)	146 (100.0)	1,2-Dichlorobenzene
146 (100.0)	**148** (58.3)	75 (50.6)	111 (49.7)	50 (36.5)	74 (32.4)	73 (18.1)	113 (16.0)	146 (100.0)	1,3-Dichlorobenzene
146 (100.0)	**148** (58.2)	111 (33.4)	75 (20.8)	74 (13.0)	113 (11.0)	150 (10.1)	50 (9.4)	146 (100.0)	1,4-Dichlorobenzene
149 (100.0)	**150** (9.0)	41 (6.3)	223 (5.7)	104 (5.0)	76 (4.8)	205 (4.5)	65 (3.4)	278	Di-n-butyl phthalate
152 (100.0)	**151** (17.5)	150 (13.5)	153 (12.6)	76 (9.3)	75 (3.8)	74 (3.7)	126 (3.6)	152 (100.0)	Acenaphthylene
222 (100.0)	**152** (78.7)	224 (62.9)	75 (20.9)	151 (17.6)	93 (15.5)	150 (13.2)	223 (13.1)		PCB-1016 (peak 1)
188 (100.0)	**152** (45.9)	190 (32.8)	153 (25.0)	76 (16.5)	189 (13.3)	151 (13.3)	75 (8.3)		PCB-1221 (peak 2)
188 (100.0)	**152** (41.8)	190 (32.5)	153 (20.6)	76 (14.7)	189 (13.2)	151 (11.2)	63 (6.8)		PCB-1221 (peak 3)
222 (100.0)	**152** (67.2)	224 (62.1)	75 (16.9)	151 (14.8)	223 (12.8)	150 (11.9)	226 (10.3)		PCB-1221 (peak 5)

Second-Peak Index—continued

Partial spectrum								M	Substance
188 (100.0)	**152 (44.8)**	190 (32.5)	153 (23.0)	76 (15.6)	151 (12.8)	189 (12.2)	63 (7.8)		PCB-1232 (peak 1)
188 (100.0)	**152 (41.4)**	190 (32.7)	153 (20.6)	76 (13.7)	189 (12.5)	151 (11.7)	63 (6.8)		PCB-1232 (peak 2)
222 (100.0)	**152 (63.4)**	224 (60.9)	75 (16.4)	151 (13.9)	223 (13.1)	150 (10.9)	93 (10.1)		PCB-1232 (peak 3)
222 (100.0)	**152 (65.0)**	224 (62.5)	75 (16.6)	151 (16.0)	223 (12.4)	150 (11.4)	93 (11.3)		PCB-1242 (peak 1)
154 (100.0)	**153 (40.8)**	152 (26.9)	155 (14.5)	76 (11.6)	151 (7.5)	51 (5.5)	77 (5.4)		PCB-1221 (peak 1)
153 (100.0)	**154 (96.5)**	152 (44.6)	76 (19.0)	151 (16.7)	155 (14.0)	63 (8.9)	150 (8.5)	154 (96.5)	Acenaphthene
159 (100.0)	**161 (89.9)**	125 (77.8)	85 (74.7)	75 (74.7)	195 (66.7)	83 (65.7)	197 (57.6)		Toxaphene (peak 3)
162 (100.0)	**164 (62.2)**	63 (45.4)	98 (31.5)	126 (13.9)	99 (12.3)	62 (12.2)	73 (11.0)	162 (100.0)	2,4-Dichlorophenol
235 (100.0)	**165 (58.4)**	237 (57.0)	75 (21.2)	199 (13.5)	88 (12.3)	236 (12.2)	82 (10.2)	318 (2.6)	4,4′-DDD
166 (100.0)	**165 (90.9)**	163 (14.2)	164 (12.6)	167 (12.4)	139 (6.5)	115 (4.2)	63 (3.9)	166 (100.0)	Fluorene
149 (100.0)	**167 (43.1)**	57 (23.3)	71 (15.4)	70 (15.0)	43 (14.3)	279 (13.7)	41 (13.6)	390	Bis(2-ethylhexyl) phthalate
195 (100.0)	**170 (79.3)**	207 (75.2)	239 (71.0)	237 (69.0)	159 (67.6)	241 (66.9)	197 (62.8)	404 (2.1)	α-Endosulfan

173 (100.0)	**171** (51.9)	175 (50.7)	79 (30.6)	91 (29.7)	93 (29.6)	81 (28.9)	94 (16.9)	250 (3.3)	Bromoform
178 (100.0)	**176** (17.7)	179 (13.9)	177 (8.6)	76 (8.2)	89 (7.6)	152 (7.2)	151 (5.6)	178 (100.0)	Anthracene
178 (100.0)	**176** (17.9)	179 (14.2)	76 (10.3)	177 (10.1)	152 (9.4)	89 (8.8)	88 (7.3)	178 (100.0)	Phenanthrene
149 (100.0)	**177** (23.7)	65 (10.6)	176 (10.5)	150 (10.2)	76 (9.6)	105 (8.5)	121 (7.0)	222 (4.6)	Diethyl phthalate
109 (100.0)	**181** (89.3)	183 (82.7)	111 (82.4)	217 (65.5)	51 (48.8)	85 (46.0)	83 (38.6)	288	β-BHC
180 (100.0)	**182** (96.3)	184 (30.2)	145 (29.0)	109 (22.1)	74 (21.6)	147 (17.1)	75 (12.8)	180 (100.0)	1,2,4-Trichlorobenzene
181 (100.0)	**183** (91.7)	219 (84.7)	217 (66.0)	109 (65.3)	111 (62.4)	51 (43.0)	85 (36.9)	288	α-BHC
181 (100.0)	**183** (98.5)	111 (96.9)	219 (83.0)	109 (74.4)	217 (58.3)	51 (55.6)	73 (55.4)	288 (2.9)	γ-BHC
181 (100.0)	**183** (94.0)	219 (89.6)	109 (85.0)	111 (74.9)	217 (64.5)	51 (55.2)	85 (49.7)	288	δ-BHC
184 (100.0)	**185** (12.3)	183 (11.5)	92 (6.8)	156 (6.4)	91 (5.9)	167 (5.1)	166 (4.9)	184 (100.0)	Benzidine
188 (100.0)	**187** (19.1)	80 (16.1)	184 (15.7)	189 (15.0)	94 (14.9)	160 (11.8)	66 (11.5)	188 (100.0)	Anthracene-d_{10} (IS)
195 (100.0)	**197** (80.5)	159 (75.6)	160 (65.9)	207 (63.4)	170 (53.7)	239 (51.2)	237 (51.2)	404 (4.9)	β-Endosulfan
196 (100.0)	**198** (97.5)	97 (56.8)	132 (42.7)	200 (33.9)	134 (28.0)	62 (25.7)	99 (22.9)	196 (100.0)	2,4,6-Trichlorophenol
202 (100.0)	**200** (20.2)	203 (18.8)	101 (15.9)	201 (13.9)	100 (11.3)	88 (8.7)	87 (5.2)	202 (100.0)	Fluoranthene

Second-Peak Index—continued

Partial spectrum								M	Substance
202 (100.0)	**200** (22.2)	201 (19.7)	203 (19.5)	101 (15.4)	100 (13.0)	88 (5.5)	199 (4.5)	202 (100.0)	Pyrene
292 (100.0)	**220** (93.4)	290 (81.5)	222 (60.7)	294 (48.6)	110 (37.9)	150 (27.9)	255 (27.8)		PCB-1248 (peak 2)
152 (100.0)	**222** (86.1)	224 (54.3)	187 (48.4)	75 (20.8)	151 (20.0)	93 (15.3)	150 (14.9)		PCB-1221 (peak 4)
186 (100.0)	**222** (91.2)	258 (79.0)	256 (78.9)	152 (54.9)	224 (53.5)	223 (38.7)	150 (35.7)		PCB-1232 (peak 4)
225 (100.0)	**223** (67.2)	227 (64.1)	190 (45.9)	188 (37.9)	118 (34.8)	260 (33.6)	262 (23.6)	258 (20.8)	Hexachlorobutadiene
228 (100.0)	**226** (26.5)	113 (23.0)	114 (22.8)	229 (20.6)	101 (12.5)	112 (10.0)	100 (9.5)	228 (100.0)	Benzo[*a*]anthracene
228 (100.0)	**226** (30.9)	113 (26.3)	114 (22.2)	229 (20.3)	227 (12.5)	112 (11.0)	100 (10.3)	228 (100.0)	Chrysene
237 (100.0)	**235** (73.0)	239 (69.7)	95 (28.4)	130 (22.6)	241 (22.2)	272 (17.8)	60 (16.9)	270 (9.7)	Hexachlorocyclopentadiene
235 (100.0)	**237** (56.6)	165 (53.8)	75 (17.6)	176 (13.3)	236 (12.5)	199 (11.9)	212 (11.0)	352	4,4′-DDT
245 (100.0)	**243** (100.0)	83 (97.3)	85 (78.4)	207 (64.9)	159 (48.6)	73 (45.9)	111 (43.2)		Toxaphene (peak 5)
252 (100.0)	**250** (20.9)	253 (20.7)	126 (14.9)	125 (10.2)	113 (7.1)	112 (6.5)	251 (6.2)	252 (100.0)	Benzo[*k*]fluoranthene
252 (100.0)	**250** (22.1)	253 (22.0)	126 (16.9)	125 (10.4)	113 (9.7)	251 (8.2)	112 (6.8)	252 (100.0)	Benzo[*a*]pyrene

248 (100.0)	**250** (88.6)	141 (69.5)	77 (58.5)	51 (37.3)	115 (29.3)	50 (24.4)	63 (22.5)	248 (100.0)	4-Bromophenyl phenyl ether
67 (100.0)	**250** (40.0)	345 (30.0)	197 (30.0)	135 (30.0)	95 (29.0)	248 (27.1)	66 (26.7)	378	Endrin aldehyde
252 (100.0)	**254** (59.1)	253 (17.8)	126 (17.3)	127 (15.5)	154 (13.7)	91 (12.6)	181 (10.8)	252 (100.0)	3,3′-Dichlorobenzidine
258 (100.0)	**256** (94.1)	186 (73.6)	260 (28.6)	150 (24.3)	188 (23.7)	75 (22.6)	93 (17.9)		PCB-1016 (peak 4)
258 (100.0)	**256** (92.9)	186 (92.9)	257 (66.1)	92 (52.7)	110 (38.8)	260 (36.2)	150 (33.0)		PCB-1016 (peak 5)
258 (100.0)	**256** (91.1)	186 (60.1)	260 (27.7)	188 (21.0)	150 (20.3)	75 (19.3)	151 (14.4)		PCB-1232 (peak 5)
258 (100.0)	**256** (92.9)	186 (63.9)	260 (28.0)	150 (21.7)	188 (20.8)	75 (20.8)	151 (15.7)		PCB-1242 (peak 3)
258 (100.0)	**256** (97.0)	186 (66.3)	260 (29.4)	150 (23.6)	188 (22.2)	75 (21.5)	151 (16.7)		PCB-1248 (peak 1)
186 (100.0)	**258** (72.6)	256 (69.3)	221 (36.1)	150 (32.7)	188 (30.9)	75 (29.1)	222 (26.9)		PCB-1016 (peak 2)
186 (100.0)	**258** (87.4)	256 (78.4)	221 (38.0)	150 (33.1)	75 (30.8)	260 (24.9)	151 (23.9)		PCB-1016 (peak 3)
186 (100.0)	**258** (84.4)	256 (80.7)	221 (37.6)	222 (35.1)	188 (34.2)	150 (31.4)	75 (29.7)		PCB-1242 (peak 2)
256 (100.0)	**258** (89.0)	186 (81.9)	188 (33.5)	260 (29.1)	150 (25.3)	75 (24.7)	290 (22.5)		PCB-1242 (peak 4)
266 (100.0)	**268** (74.3)	264 (70.0)	165 (45.5)	95 (39.5)	167 (38.7)	60 (26.5)	270 (26.1)	264 (70.0)	Pentachlorophenol
100 (100.0)	**272** (54.9)	274 (44.7)	102 (38.1)	65 (35.0)	270 (33.2)	237 (20.4)	276 (17.3)		Chlordane (peak 2)

Second-Peak Index—continued

Partial spectrum								*M*	Substance
100 (100.0)	**272** (52.8)	65 (47.8)	274 (45.0)	102 (34.1)	270 (28.5)	237 (21.6)	135 (17.7)	370 (2.2)	Heptachlor
272 (100.0)	**274** (97.7)	229 (71.6)	270 (62.5)	227 (62.5)	237 (56.8)	387 (48.9)	239 (48.9)	420 (5.7)	Endosulfan sulfate
276 (100.0)	**277** (23.7)	274 (21.3)	138 (19.4)	137 (17.8)	275 (11.4)	136 (8.9)	272 (5.2)	276 (100.0)	Benzo[*ghi*]perylene
276 (100.0)	**277** (24.1)	274 (21.5)	138 (18.2)	137 (13.9)	275 (9.2)	136 (7.8)	125 (5.2)	276 (100.0)	Indeno[1,2,3-*cd*]pyrene
149 (100.0)	**279** (10.7)	150 (9.7)	43 (7.6)	41 (7.5)	57 (5.5)	71 (3.8)	69 (3.2)	390	Di-n-octyl phthalate
284 (100.0)	**286** (83.7)	282 (50.7)	142 (35.3)	288 (32.8)	249 (28.1)	107 (25.6)	144 (22.6)	282 (50.7)	Hexachlorobenzene
292 (100.0)	**290** (92.2)	220 (61.0)	294 (53.1)	222 (40.2)	110 (27.3)	150 (19.1)	111 (15.6)		PCB-1242 (peak 5)
292 (100.0)	**290** (90.5)	220 (64.9)	294 (50.8)	222 (37.2)	110 (22.5)	150 (20.0)	111 (17.2)		PCB-1248 (peak 4)
292 (100.0)	**290** (88.0)	220 (62.9)	294 (50.0)	222 (38.3)	110 (29.1)	150 (18.6)	111 (16.7)		PCB-1248 (peak 5)
292 (100.0)	**290** (90.9)	220 (63.6)	294 (56.2)	222 (40.2)	110 (31.3)	326 (23.7)	184 (21.2)		PCB-1254 (peak 1)
220 (100.0)	**292** (88.8)	290 (70.2)	222 (67.0)	255 (45.7)	110 (45.7)	257 (43.6)	294 (39.9)		PCB-1248 (peak 3)
232 (100.0)	**303** (94.8)	230 (94.8)	169 (75.3)	305 (66.0)	171 (64.9)	196 (48.5)	231 (46.4)		Chlordane (peak 3)

246 (100.0)	**318** (76.8)	248 (72.9)	316 (57.1)	176 (57.0)	105 (40.4)	320 (35.0)	75 (23.8)	316 (57.1)	4,4′-DDE
322 (100.0)	**320** (78.5)	324 (47.9)	74 (37.4)	97 (36.0)	259 (34.3)	257 (32.9)	194 (31.6)	320 (78.5)	2,3,7,8-Tetrachlorodibenzo-*p*-dioxin
360 (100.0)	**326** (91.7)	362 (78.7)	328 (68.1)	324 (54.2)	145 (47.8)	288 (47.3)	254 (44.4)		PCB-1254 (peak 5)
326 (100.0)	**328** (71.0)	254 (63.8)	324 (62.1)	256 (58.8)	127 (33.1)	128 (31.6)	184 (26.7)		PCB-1254 (peak 2)
326 (100.0)	**328** (76.4)	324 (66.4)	254 (41.5)	256 (31.2)	330 (20.9)	128 (18.8)	184 (16.7)		PCB-1254 (peak 3)
326 (100.0)	**328** (73.1)	324 (62.3)	254 (40.2)	256 (35.8)	127 (27.9)	128 (22.0)	109 (23.4)		PCB-1254 (peak 4)
81 (100.0)	**353** (38.6)	355 (35.9)	53 (25.0)	351 (24.6)	51 (20.7)	357 (16.2)	61 (15.7)	386 (2.3)	Heptachlor epoxide
360 (100.0)	**362** (73.5)	290 (60.3)	288 (42.5)	145 (41.3)	358 (37.7)	146 (27.4)	364 (21.1)		PCB-1260 (peak 1)
360 (100.0)	**362** (76.4)	290 (62.7)	288 (46.5)	358 (42.4)	145 (39.7)	144 (32.2)	364 (30.6)		PCB-1260 (peak 2)
360 (100.0)	**362** (75.5)	290 (59.4)	288 (43.1)	358 (42.8)	145 (37.8)	364 (30.3)	144 (26.1)		PCB-1260 (peak 3)
373 (100.0)	**375** (91.8)	371 (44.8)	377 (44.4)	237 (28.9)	272 (26.3)	239 (24.1)	65 (23.7)		Chlordane (peak 4)
373 (100.0)	**375** (83.9)	377 (43.5)	371 (40.5)	66 (39.3)	237 (29.8)	272 (25.6)	65 (25.6)		Chlordane (peak 5)
394 (100.0)	**396** (96.6)	324 (66.5)	162 (57.1)	398 (49.2)	322 (47.0)	392 (46.6)	326 (45.5)		PCB-1260 (peak 4)
394 (100.0)	**396** (96.7)	324 (66.2)	162 (59.3)	398 (49.2)	322 (46.3)	392 (46.1)	326 (45.6)		PCB-1260 (peak 5)

Third-Peak Index

Partial spectrum								M	Substance
53 (100.0)	52 (79.6)	**26 (78.5)**	51 (33.5)	27 (16.0)	25 (10.2)	38 (9.8)	50 (9.1)	53 (100.0)	Acrylonitrile
64 (100.0)	28 (84.9)	**27 (82.6)**	29 (78.1)	26 (36.9)	66 (35.0)	49 (27.0)	51 (8.8)	64 (100.0)	Chloroethane
55 (100.0)	41 (38.2)	**27 (31.2)**	54 (28.6)	90 (26.5)	39 (20.2)	62 (19.5)	49 (17.0)	126	1,4-Dichlorobutane (IS)
41 (100.0)	77 (93.2)	**39 (54.6)**	79 (36.4)	27 (20.5)	49 (14.0)	38 (13.4)	37 (7.9)	156	2-Bromo-1-chloropropane (IS)
45 (100.0)	121 (69.1)	**41 (64.1)**	77 (26.4)	123 (24.0)	79 (18.3)	49 (11.1)	107 (10.7)	170	Bis(2-chloroisopropyl) ether
149 (100.0)	150 (9.0)	**41 (6.3)**	223 (5.7)	104 (5.0)	76 (4.8)	205 (4.5)	65 (3.4)	278	Di-n-butyl phthalate
63 (100.0)	62 (79.2)	**41 (70.1)**	76 (61.7)	39 (54.9)	27 (44.7)	65 (30.8)	64 (27.0)	112 (5.8)	1,2-Dichloropropane
74 (100.0)	42 (54.1)	**43 (24.9)**	40 (3.8)	41 (3.1)	75 (3.0)	44 (2.8)	–	74 (100.0)	*N*-Nitrosodimethylamine
63 (100.0)	43 (98.1)	**44 (71.5)**	65 (32.3)	42 (25.0)	62 (24.7)	106 (24.5)	57 (20.2)	106 (24.5)	2-Chloroethyl vinyl ether
83 (100.0)	85 (66.5)	**47 (20.8)**	48 (12.8)	79 (12.4)	81 (11.9)	129 (10.1)	87 (9.4)	162	Bromodichloromethane
83 (100.0)	85 (64.7)	**47 (29.7)**	35 (14.4)	48 (11.9)	49 (10.5)	87 (9.5)	37 (4.8)	118	Chloroform
50 (100.0)	52 (34.2)	**49 (13.3)**	47 (11.9)	35 (8.4)	48 (5.0)	51 (4.6)	37 (3.0)	50 (100.0)	Chloromethane

78 (100.0)	77 (25.9)	**50** (20.8)	51 (20.6)	52 (18.8)	39 (9.3)	79 (6.4)	74 (5.8)	78 (100.0)	Benzene
85 (100.0)	87 (30.9)	**50** (24.6)	35 (22.0)	31 (16.3)	47 (8.7)	101 (7.6)	66 (6.8)	120	Dichlorodifluoromethane
77 (100.0)	123 (75.6)	**51** (43.8)	50 (16.0)	93 (14.8)	65 (14.0)	74 (7.9)	78 (6.7)	123 (75.6)	Nitrobenzene
184 (100.0)	63 (67.9)	**53** (61.2)	91 (39.3)	62 (35.7)	107 (33.9)	154 (31.3)	79 (31.3)	184 (100.0)	2,4-Dinitrophenol
56 (100.0)	27 (66.2)	**55** (64.9)	26 (44.5)	29 (37.7)	37 (11.3)	38 (9.3)	53 (8.6)	56 (100.0)	Acrolein
149 (100.0)	167 (43.1)	**57** (23.3)	71 (15.4)	70 (15.0)	43 (14.3)	279 (13.7)	41 (13.6)	390	Bis(2-ethylhexyl) phthalate
97 (100.0)	99 (60.9)	**61** (51.2)	63 (16.3)	26 (11.8)	117 (10.7)	119 (10.2)	101 (10.2)	132	1,1,1-Trichloroethane
61 (100.0)	96 (44.3)	**63** (31.9)	98 (28.9)	60 (18.9)	26 (13.4)	35 (9.2)	62 (8.4)	96 (44.3)	1,1-Dichloroethylene
162 (100.0)	164 (62.2)	**63** (45.4)	98 (31.5)	126 (13.9)	99 (12.3)	62 (12.2)	73 (11.0)	162 (100.0)	2,4-Dichlorophenol
165 (100.0)	89 (64.4)	**63** (53.0)	90 (22.7)	119 (18.5)	78 (18.4)	51 (18.1)	153 (3.4)	182 (9.1)	2,4-Dinitrotoluene
139 (100.0)	65 (68.0)	**63** (59.2)	81 (45.2)	53 (34.8)	64 (34.2)	109 (26.7)	51 (19.0)	139 (100.0)	2-Nitrophenol
128 (100.0)	130 (31.2)	**64** (29.3)	63 (16.1)	92 (12.3)	129 (6.7)	73 (5.0)	65 (4.8)	128 (100.0)	2-Chlorophenol
62 (100.0)	27 (39.8)	**64** (31.1)	49 (24.9)	26 (13.1)	63 (12.5)	61 (8.8)	51 (7.8)	98 (7.3)	1,2-Dichloroethane
62 (100.0)	27 (76.4)	**64** (30.8)	26 (24.2)	25 (10.3)	61 (8.4)	35 (7.3)	60 (6.5)	62 (100.0)	Vinyl chloride

Third-Peak Index—continued

Partial spectrum								*M*	Substance
63 (100.0)	27 (42.8)	65 (31.4)	26 (13.2)	83 (12.9)	61 (8.9)	62 (8.4)	85 (8.3)	98 (6.8)	1,1-Dichloroethane
149 (100.0)	177 (23.7)	65 (10.6)	176 (10.5)	150 (10.2)	76 (9.6)	105 (8.5)	121 (7.0)	222 (4.6)	Diethyl phthalate
91 (100.0)	106 (27.3)	65 (10.9)	51 (10.7)	77 (9.2)	78 (8.0)	92 (7.6)	39 (6.5)	106 (27.3)	Ethylbenzene
100 (100.0)	272 (52.8)	65 (47.8)	274 (45.0)	102 (34.1)	270 (28.5)	237 (21.6)	135 (17.7)	370 (2.2)	Heptachlor
94 (100.0)	66 (32.3)	65 (22.0)	95 (6.6)	63 (5.6)	55 (4.7)	50 (3.9)	51 (3.5)	94 (100.0)	Phenol
91 (100.0)	92 (57.8)	65 (14.4)	39 (12.8)	63 (10.4)	51 (8.3)	50 (5.9)	62 (5.0)	92 (57.8)	Toluene
101 (100.0)	103 (59.1)	66 (17.6)	35 (12.8)	105 (11.0)	47 (10.9)	31 (9.3)	68 (6.0)	136	Trichlorofluoromethane
146 (100.0)	148 (60.1)	75 (52.5)	111 (51.5)	50 (37.5)	74 (31.8)	113 (18.0)	73 (15.9)	146 (100.0)	1,2-Dichlorobenzene
146 (100.0)	148 (58.3)	75 (50.6)	111 (49.7)	50 (36.5)	74 (32.4)	73 (18.1)	113 (16.0)	146 (100.0)	1,3-Dichlorobenzene
107 (100.0)	142 (76.4)	77 (43.7)	144 (24.8)	51 (17.4)	78 (12.1)	79 (11.8)	143 (9.0)	142 (76.4)	*p*-Chloro-*m*-cresol
75 (100.0)	39 (50.3)	77 (28.9)	49 (21.2)	110 (19.6)	38 (13.9)	112 (12.2)	37 (12.1)	110 (19.6)	*cis*-1,3-Dichloropropene
75 (100.0)	39 (55.0)	77 (29.1)	49 (20.4)	110 (19.8)	38 (14.4)	37 (12.8)	112 (11.7)	110 (19.8)	*trans*-1,3-Dichloropropene

66 (100.0)	91 (42.7)	79 (38.9)	65 (31.9)	101 (29.6)	263 (29.4)	265 (22.3)	261 (22.3)	362	Aldrin
188 (100.0)	187 (19.1)	80 (16.1)	184 (15.7)	189 (15.0)	94 (14.9)	160 (11.8)	66 (11.5)	188 (100.0)	Anthracene-d_{10} (IS)
79 (100.0)	49 (37.9)	81 (33.1)	51 (11.9)	29 (9.5)	78 (5.5)	80 (3.8)	48 (2.9)	114	Bis(chloromethyl) ether
79 (100.0)	82 (34.8)	81 (34.7)	77 (22.6)	108 (15.4)	53 (13.6)	107 (11.5)	263 (10.1)	378	Dieldrin
245 (100.0)	243 (100.0)	83 (97.3)	85 (78.4)	207 (64.9)	159 (48.6)	73 (45.9)	111 (43.2)		Toxaphene (peak 5)
49 (100.0)	84 (60.1)	86 (38.3)	51 (28.4)	47 (22.0)	35 (14.6)	48 (9.6)	28 (6.6)	84 (60.1)	Methylene chloride
165 (100.0)	63 (47.7)	89 (41.7)	90 (30.4)	77 (23.9)	51 (22.2)	64 (21.0)	78 (20.8)	182 (2.5)	2,6-Dinitrotoluene
94 (100.0)	96 (89.3)	93 (21.6)	79 (21.3)	81 (20.5)	95 (14.1)	91 (9.0)	92 (4.3)	94 (100.0)	Bromomethane
93 (100.0)	63 (68.7)	95 (30.4)	123 (24.9)	65 (22.1)	125 (7.2)	106 (4.9)	49 (4.5)	172	Bis(2-chloroethoxy)methane
93 (100.0)	63 (57.8)	95 (34.4)	65 (16.6)	49 (5.4)	94 (3.8)	106 (3.0)	142 (2.9)	142 (2.9)	Bis(2-chloroethyl) ether
83 (100.0)	85 (66.3)	95 (15.6)	60 (14.1)	61 (11.9)	96 (11.1)	168 (9.9)	87 (9.8)	166 (7.7)	1,1,2,2-Tetrachloroethane
130 (100.0)	132 (98.5)	95 (84.3)	97 (55.1)	60 (37.4)	134 (30.2)	62 (13.1)	35 (10.2)	130 (100.0)	Trichloroethylene
129 (100.0)	127 (76.2)	97 (39.6)	83 (36.0)	61 (35.2)	79 (24.1)	131 (23.7)	81 (23.4)	206 (2.6)	Dibromochloromethane
196 (100.0)	198 (97.5)	97 (56.8)	132 (42.7)	200 (33.9)	134 (28.0)	62 (25.7)	99 (22.9)	196 (100.0)	2,4,6-Trichlorophenol

Third-Peak Index—continued

Partial spectrum								*M*	Substance
61 (100.0)	96 (60.2)	**98** (39.3)	63 (32.3)	60 (29.7)	26 (16.3)	62 (12.2)	25 (8.6)	96 (60.2)	*trans*-1,2-Dichloroethylene
97 (100.0)	83 (86.0)	**99** (61.7)	61 (54.7)	85 (53.0)	96 (28.3)	98 (20.6)	63 (17.7)	132 (15.7)	1,1,2-Trichloroethane
85 (100.0)	83 (95.7)	**100** (82.9)	73 (71.4)	329 (58.6)	159 (57.1)	51 (55.7)	75 (52.9)		Toxaphene (peak 4)
198 (100.0)	51 (52.3)	**105** (44.5)	53 (41.9)	121 (41.5)	77 (24.7)	52 (24.2)	50 (24.2)	198 (100.0)	4,6-Dinitro-*o*-cresol
65 (100.0)	139 (83.6)	**109** (43.4)	53 (29.7)	81 (27.6)	63 (27.3)	93 (18.9)	62 (14.8)	139 (83.6)	4-Nitrophenol
100 (100.0)	99 (46.5)	**109** (41.9)	102 (39.5)	197 (34.9)	199 (32.6)	125 (30.2)	77 (27.9)		Toxaphene (peak 2)
181 (100.0)	183 (98.5)	**111** (96.9)	219 (83.0)	109 (74.4)	217 (58.3)	51 (55.6)	73 (55.4)	288 (2.9)	γ-BHC
146 (100.0)	148 (58.2)	**111** (33.4)	75 (20.8)	74 (13.0)	113 (11.0)	150 (10.1)	50 (9.4)	146 (100.0)	1,4-Dichlorobenzene
228 (100.0)	226 (26.5)	**113** (23.0)	114 (22.8)	229 (20.6)	101 (12.5)	112 (10.0)	100 (9.5)	228 (100.0)	Benzo[*a*]anthracene
228 (100.0)	226 (30.9)	**113** (26.3)	114 (22.2)	229 (20.3)	227 (12.5)	112 (11.0)	100 (10.3)	228 (100.0)	Chrysene
112 (100.0)	77 (67.6)	**114** (31.4)	51 (25.8)	50 (25.0)	45 (11.9)	74 (11.4)	75 (9.7)	112 (100.0)	Chlorobenzene
117 (100.0)	119 (93.7)	**121** (30.7)	82 (28.3)	47 (25.6)	35 (18.7)	84 (18.3)	49 (8.9)	152	Carbon tetrachloride

107 (100.0)	122 (92.1)	**121** **(49.6)**	77 (26.9)	91 (19.4)	79 (15.0)	78 (9.0)	51 (8.9)	122 (92.1)	2,4-Dimethylphenol
159 (100.0)	161 (89.9)	**125** **(77.8)**	85 (74.7)	75 (74.7)	195 (66.7)	83 (65.7)	197 (57.6)		Toxaphene (peak 3)
49 (100.0)	130 (38.4)	**128** **(31.1)**	51 (28.4)	79 (26.2)	81 (25.6)	93 (22.4)	95 (16.2)	128 (31.1)	Bromochloromethane (IS)
128 (100.0)	127 (14.0)	**129** **(11.4)**	102 (8.9)	126 (7.8)	51 (5.8)	75 (4.9)	63 (4.7)	128 (100.0)	Naphthalene
70 (100.0)	43 (70.7)	**130** **(44.8)**	42 (41.4)	41 (33.9)	101 (16.6)	58 (15.8)	113 (10.8)	130 (44.8)	*N*-Nitrosodi-n-propylamine
278 (100.0)	139 (38.6)	**138** **(27.2)**	279 (23.4)	276 (21.9)	137 (16.3)	125 (15.2)	113 (8.8)	278 (100.0)	Dibenzo[*a*,*h*]anthracene
82 (100.0)	54 (14.8)	**138** **(12.2)**	41 (7.7)	53 (6.9)	83 (5.3)	67 (5.3)	55 (5.3)	138 (12.2)	Isophorone
248 (100.0)	250 (88.6)	**141** **(69.5)**	77 (58.5)	51 (37.3)	115 (29.3)	50 (24.4)	63 (22.5)	248 (100.0)	4-Bromophenyl phenyl ether
152 (100.0)	151 (17.5)	**150** **(13.5)**	153 (12.6)	76 (9.3)	75 (3.8)	74 (3.7)	126 (3.6)	152 (100.0)	Acenaphthylene
149 (100.0)	279 (10.7)	**150** **(9.7)**	43 (7.6)	41 (7.5)	57 (5.5)	71 (3.8)	69 (3.2)	390	Di-n-octyl phthalate
153 (100.0)	154 (96.5)	**152** **(44.6)**	76 (19.0)	151 (16.7)	155 (14.0)	63 (8.9)	150 (8.5)	154 (96.5)	Acenaphthene
154 (100.0)	153 (40.8)	**152** **(26.9)**	155 (14.5)	76 (11.6)	151 (7.5)	51 (5.5)	77 (5.4)		PCB-1221 (peak 1)
195 (100.0)	197 (80.5)	**159** **(75.6)**	160 (65.9)	207 (63.4)	170 (53.7)	239 (51.2)	237 (51.2)	404 (4.9)	β-Endosulfan
166 (100.0)	165 (90.9)	**163** **(14.2)**	164 (12.6)	167 (12.4)	139 (6.5)	115 (4.2)	63 (3.9)	166 (100.0)	Fluorene

Third-Peak Index—continued

Partial spectrum								*M*	Substance
162 (100.0)	127 (32.1)	**164** (29.0)	126 (17.3)	163 (10.9)	63 (8.3)	75 (7.6)	81 (6.7)	162 (100.0)	2-Chloronaphthalene
166 (100.0)	129 (77.1)	**164** (74.2)	131 (73.4)	168 (54.3)	94 (53.8)	47 (35.7)	96 (33.6)	164 (74.2)	1,1,2,2-Tetrachloroethene
235 (100.0)	237 (56.6)	**165** (53.8)	75 (17.6)	176 (13.3)	236 (12.5)	199 (11.9)	212 (11.0)	352	4,4′-DDT
173 (100.0)	171 (51.9)	**175** (50.7)	79 (30.6)	91 (29.7)	93 (29.6)	81 (28.9)	94 (16.9)	250 (3.3)	Bromoform
178 (100.0)	176 (17.7)	**179** (13.9)	177 (8.6)	76 (8.2)	89 (7.6)	152 (7.2)	151 (5.6)	178 (100.0)	Anthracene
178 (100.0)	176 (17.9)	**179** (14.2)	76 (10.3)	177 (10.1)	152 (9.4)	89 (8.8)	88 (7.3)	178 (100.0)	Phenanthrene
77 (100.0)	51 (37.6)	**182** (24.2)	105 (15.5)	78 (11.3)	50 (9.5)	152 (5.7)	153 (3.4)	184	1,2-Diphenylhydrazine [azobenzene]
77 (100.0)	51 (23.7)	**182** (21.9)	105 (18.4)	78 (7.3)	152 (5.3)	50 (4.6)	183 (2.9)	198	*N*-Nitrosodiphenylamine [diphenylamine]
184 (100.0)	185 (12.3)	**183** (11.5)	92 (6.8)	156 (6.4)	91 (5.9)	167 (5.1)	166 (4.9)	184 (100.0)	Benzidine
109 (100.0)	181 (89.3)	**183** (82.7)	111 (82.4)	217 (65.5)	51 (48.8)	85 (46.0)	83 (38.6)	288	β-BHC
180 (100.0)	182 (96.3)	**184** (30.2)	145 (29.0)	109 (22.1)	74 (21.6)	147 (17.1)	75 (12.8)	180 (100.0)	1,2,4-Trichlorobenzene
258 (100.0)	256 (94.1)	**186** (73.6)	260 (28.6)	150 (24.3)	188 (23.7)	75 (22.6)	93 (17.9)		PCB-1016 (peak 4)

258 (100.0)	256 (92.9)	**186** (92.9)	257 (66.1)	92 (52.7)	110 (38.8)	260 (36.2)	150 (33.0)		PCB-1016 (peak 5)
258 (100.0)	256 (91.1)	**186** (92.9)	260 (27.7)	188 (21.0)	150 (20.3)	75 (19.3)	151 (14.4)		PCB-1232 (peak 5)
158 (100.0)	256 (92.9)	**186** (63.9)	260 (28.0)	150 (21.7)	188 (20.8)	75 (20.8)	151 (15.7)		PCB-1242 (peak 3)
256 (100.0)	258 (89.0)	**186** (81.9)	188 (33.5)	260 (29.1)	150 (25.3)	75 (24.7)	290 (22.5)		PCB-1242 (peak 4)
258 (100.0)	256 (97.0)	**186** (66.3)	260 (29.4)	150 (23.6)	188 (22.2)	75 (21.5)	151 (16.7)		PCB-1248 (peak 1)
188 (100.0)	152 (45.9)	**190** (32.8)	153 (25.0)	76 (16.5)	189 (13.3)	151 (13.3)	75 (8.3)		PCB-1221 (peak 2)
188 (100.0)	152 (41.8)	**190** (32.5)	153 (20.6)	76 (14.7)	189 (13.2)	151 (11.2)	63 (6.8)		PCB-1221 (peak 3)
188 (100.0)	152 (44.8)	**190** (32.5)	153 (23.0)	76 (15.6)	151 (12.8)	189 (12.2)	63 (7.8)		PCB-1232 (peak 1)
188 (100.0)	152 (41.4)	**190** (32.7)	153 (20.6)	76 (13.7)	189 (12.5)	151 (11.7)	63 (6.8)		PCB-1232 (peak 2)
163 (100.0)	77 (17.7)	**194** (10.2)	164 (9.5)	92 (8.0)	76 (7.6)	50 (7.2)	133 (6.2)	194 (10.2)	Dimethyl phthalate
159 (100.0)	125 (100.0)	**197** (80.4)	51 (79.4)	75 (78.4)	161 (73.2)	209 (68.0)	233 (62.9)		Toxaphene (peak 1)
117 (100.0)	119 (83.0)	**201** (57.8)	203 (37.5)	199 (34.7)	166 (32.3)	94 (30.5)	47 (30.4)	234	Hexachloroethane
202 (100.0)	200 (22.2)	**201** (19.7)	203 (19.5)	101 (15.4)	100 (13.0)	88 (5.5)	199 (4.5)	202 (100.0)	Pyrene
202 (100.0)	200 (20.2)	**203** (18.8)	101 (15.9)	201 (13.9)	100 (11.3)	88 (8.7)	87 (5.2)	202 (100.0)	Fluoranthene

Third-Peak Index—continued

Partial spectrum								*M*	Substance
149 (100.0)	91 (69.5)	**206 (24.4)**	65 (16.4)	123 (14.9)	104 (14.2)	132 (12.7)	150 (11.2)	312 (2.0)	Butyl benzyl phthalate
204 (100.0)	141 (50.0)	**206 (36.9)**	77 (28.7)	51 (20.3)	115 (12.7)	75 (12.7)	205 (12.0)	204 (100.0)	4-Chlorophenyl phenyl ether
195 (100.0)	170 (79.3)	**207 (75.2)**	239 (71.0)	237 (69.0)	159 (67.6)	241 (66.9)	197 (62.8)	404 (2.1)	α-Endosulfan
181 (100.0)	183 (91.7)	**219 (84.7)**	217 (66.0)	109 (65.3)	111 (62.4)	51 (43.0)	85 (36.9)	288	α-BHC
181 (100.0)	183 (94.0)	**219 (89.6)**	109 (85.0)	111 (74.9)	217 (64.5)	51 (55.2)	85 (49.7)	288	δ-BHC
292 (100.0)	290 (92.2)	**220 (61.0)**	294 (53.1)	222 (40.2)	110 (27.3)	150 (19.1)	111 (15.6)		PCB-1242 (peak 5)
292 (100.0)	290 (90.5)	**220 (64.9)**	294 (50.8)	222 (37.2)	110 (22.5)	150 (20.0)	111 (17.2)		PCB-1248 (peak 4)
292 (100.0)	290 (88.8)	**220 (62.9)**	294 (50.0)	222 (38.3)	110 (29.1)	150 (18.6)	111 (16.7)		PCB-1248 (peak 5)
292 (100.0)	290 (90.9)	**220 (63.6)**	294 (56.2)	222 (40.2)	110 (31.3)	326 (23.7)	184 (21.2)		PCB-1254 (peak 1)
222 (100.0)	152 (78.7)	**224 (62.9)**	75 (20.9)	151 (17.6)	93 (15.5)	150 (13.2)	223 (13.1)		PCB-1016 (peak 1)
152 (100.0)	222 (86.1)	**224 (54.3)**	187 (48.4)	75 (20.8)	151 (20.0)	93 (15.3)	150 (14.9)		PCB-1221 (peak 4)
222 (100.0)	152 (67.2)	**224 (62.1)**	75 (16.9)	151 (14.8)	223 (12.8)	150 (11.9)	226 (10.3)		PCB-1221 (peak 5)

222 (100.0)	152 (63.4)	**224** (60.9)	75 (16.4)	151 (13.9)	223 (13.1)	150 (10.9)	93 (10.1)		PCB-1232 (peak 3)
222 (100.0)	152 (65.0)	**224** (62.5)	75 (16.6)	151 (16.0)	223 (12.4)	150 (11.4)	93 (11.3)		PCB-1242 (peak 1)
225 (100.0)	223 (67.2)	**227** (64.1)	190 (45.9)	188 (37.9)	118 (34.8)	260 (33.6)	262 (23.6)	258 (20.8)	Hexachlorobutadiene
272 (100.0)	274 (97.7)	**229** (71.6)	270 (62.5)	227 (62.5)	237 (56.8)	387 (48.9)	239 (48.9)	420 (5.7)	Endosulfan sulfate
232 (100.0)	303 (94.8)	**230** (94.8)	169 (75.3)	305 (66.0)	171 (64.9)	196 (48.5)	231 (46.4)		Chlordane (peak 3)
235 (100.0)	165 (58.4)	**237** (57.0)	75 (21.2)	199 (13.5)	88 (12.3)	236 (12.2)	82 (10.2)	318 (2.6)	4,4′-DDD
100 (100.0)	102 (35.2)	**238** (33.0)	65 (30.4)	101 (23.7)	236 (19.3)	240 (17.4)	66 (14.4)		Chlordane (peak 1)
237 (100.0)	235 (73.0)	**239** (69.7)	95 (28.4)	130 (22.6)	241 (22.2)	272 (17.8)	60 (16.9)	270 (9.7)	Hexachlorocyclopentadiene
246 (100.0)	318 (76.8)	**248** (72.9)	316 (57.1)	176 (57.0)	105 (40.4)	320 (35.0)	75 (23.8)	316 (57.1)	4,4′-DDE
252 (100.0)	126 (23.0)	**250** (21.7)	253 (20.5)	125 (14.7)	113 (10.8)	124 (9.7)	112 (8.8)	252 (100.0)	Benzo[*b*]fluoranthene
252 (100.0)	250 (20.9)	**253** (20.7)	126 (14.9)	125 (10.2)	113 (7.1)	112 (6.5)	251 (6.2)	252 (100.0)	Benzo[*k*]fluoranthene
252 (100.0)	250 (22.1)	**253** (22.0)	126 (16.9)	125 (10.4)	113 (9.7)	251 (8.2)	112 (6.8)	252 (100.0)	Benzo[*a*]pyrene
252 (100.0)	254 (59.1)	**253** (17.8)	126 (17.3)	127 (15.5)	154 (13.7)	91 (12.6)	181 (10.8)	252 (100.0)	3,3′-Dichlorobenzidine
326 (100.0)	328 (71.0)	**254** (63.8)	324 (62.1)	256 (58.8)	127 (33.1)	128 (31.6)	184 (26.7)		PCB-1254 (peak 2)

Third-Peak Index—continued

Partial spectrum								M	Substance
186 (100.0)	258 (72.6)	**256** (69.3)	221 (36.1)	150 (32.7)	188 (30.9)	75 (29.1)	222 (26.9)		PCB-1016 (peak 2)
186 (100.0)	258 (87.4)	**256** (78.4)	221 (38.0)	150 (33.1)	75 (30.8)	260 (24.9)	151 (23.9)		PCB-1016 (peak 3)
186 (100.0)	258 (84.4)	**256** (80.7)	221 (37.6)	222 (35.1)	188 (34.2)	150 (31.4)	75 (29.7)		PCB-1242 (peak 2)
186 (100.0)	222 (91.2)	**258** (79.0)	256 (78.9)	152 (54.9)	224 (53.5)	223 (38.7)	150 (35.7)		PCB-1232 (peak 4)
266 (100.0)	268 (74.3)	**264** (70.0)	165 (45.5)	95 (39.5)	167 (38.7)	60 (26.5)	270 (26.1)	264 (70.0)	Pentachlorophenol
276 (100.0)	277 (23.7)	**274** (21.3)	138 (19.4)	137 (17.8)	275 (11.4)	136 (8.9)	272 (5.2)	276 (100.0)	Benzo[*ghi*]perylene
100 (100.0)	272 (54.9)	**274** (44.7)	102 (38.1)	65 (35.0)	270 (33.2)	237 (20.4)	276 (17.3)		Chlordane (peak 2)
276 (100.0)	277 (24.1)	**274** (21.5)	138 (18.2)	137 (13.9)	275 (9.2)	136 (7.8)	125 (5.2)	276 (100.0)	Indeno[1,2,3-*cd*]pyrene
284 (100.0)	286 (83.7)	**282** (50.7)	142 (35.3)	288 (32.8)	249 (28.1)	107 (25.6)	144 (22.6)	282 (50.7)	Hexachlorobenzene
292 (100.0)	220 (93.4)	**290** (81.5)	222 (60.7)	294 (48.6)	110 (37.9)	150 (27.9)	255 (27.8)		PCB-1248 (peak 2)
220 (100.0)	292 (88.8)	**290** (70.2)	222 (67.0)	255 (45.7)	110 (45.7)	257 (43.6)	294 (39.9)		PCB-1248 (peak 3)
360 (100.0)	362 (73.5)	**290** (60.3)	288 (42.5)	145 (41.3)	358 (37.7)	146 (27.4)	364 (27.1)		PCB-1260 (peak 1)

360 (100.0)	362 (76.4)	**290** (62.7)	288 (46.5)	358 (42.4)	145 (39.7)	144 (32.2)	364 (30.6)		PCB-1260 (peak 2)
360 (100.0)	362 (75.5)	**290** (59.4)	288 (43.1)	358 (42.8)	145 (37.8)	364 (30.3)	144 (26.1)		PCB-1260 (peak 3)
67 (100.0)	83 (55.6)	**317** (50.8)	85 (45.7)	250 (36.1)	345 (35.6)	319 (33.7)	315 (33.3)	378 (2.8)	Endrin
326 (100.0)	328 (76.4)	**324** (66.4)	254 (41.5)	256 (31.2)	330 (20.9)	128 (18.8)	184 (16.7)		PCB-1254 (peak 3)
326 (100.0)	328 (73.1)	**324** (62.3)	254 (40.2)	256 (35.8)	127 (27.9)	128 (22.0)	109 (23.4)		PCB-1254 (peak 4)
394 (100.0)	396 (96.6)	**324** (66.5)	162 (57.1)	398 (49.2)	322 (47.0)	392 (46.6)	326 (45.5)		PCB-1260 (peak 4)
394 (100.0)	396 (96.7)	**324** (66.2)	162 (59.3)	398 (49.2)	322 (46.3)	392 (46.1)	326 (45.6)		PCB-1260 (peak 5)
323 (100.0)	320 (78.5)	**324** (47.9)	74 (37.4)	97 (36.0)	259 (34.3)	257 (32.9)	194 (31.6)	320 (78.5)	2,3,7,8-Tetrachlordibenzo-*p*-dioxin
67 (100.0)	250 (40.0)	**345** (30.0)	197 (30.0)	135 (30.0)	95 (29.0)	248 (27.1)	66 (26.7)	378	Endrin aldehyde
81 (100.0)	353 (38.6)	**355** (35.9)	53 (25.0)	351 (24.6)	51 (20.7)	357 (16.2)	61 (15.7)	386 (2.3)	Heptachlor epoxide
360 (100.0)	326 (91.7)	**362** (78.7)	328 (68.1)	324 (54.2)	145 (47.8)	288 (47.3)	254 (44.4)		PCB-1254 (peak 5)
373 (100.0)	375 (91.8)	**371** (44.8)	377 (44.4)	237 (28.9)	272 (26.3)	239 (24.1)	65 (23.7)		Chlordane (peak 4)
373 (100.0)	375 (83.9)	**377** (43.5)	371 (40.5)	66 (39.3)	237 (29.8)	272 (25.6)	65 (25.6)		Chlordane (peak 5)

SYNONYMS INDEX

SYNONYMS INDEX

SYNONYMS INDEX